THE STORY

OF

THE TELEGRAPH.

Cyrus W. Field

NEW YORK, RUDD & CARLETON.

"Their line is gone out through all the earth,
And their words to the end of the world."
Psalms xix. 4

THE STORY OF THE TELEGRAPH, AND A HISTORY OF THE GREAT ATLANTIC CABLE;

A COMPLETE RECORD OF THE INCEPTION, PROGRESS, AND FINAL SUCCESS OF THAT UNDERTAKING. A GENERAL HISTORY OF LAND AND OCEANIC TELEGRAPHS. DESCRIPTIONS OF TELEGRAPHIC APPARATUS, AND BIOGRAPHICAL SKETCHES OF THE PRINCIPAL PERSONS CONNECTED WITH THE GREAT WORK.

BY
CHARLES F. BRIGGS,
AND
AUGUSTUS MAVERICK.

Abundantly and Beautifully Illustrated.

NEW YORK:
RUDD & CARLETON, 310 BROADWAY.
M DCCC LVIII.

R. CRAIGHEAD,
Printer, Stereotyper, and Electrotyper,
Carton Building,
81, 83, *and* 85 *Centre Street.*

"What hath God Wrought!"

First Message over Morse's line—May 27, 1844.

"The Queen is convinced that the President will join with her in fervently hoping that the Electric Cable which now connects Great Britain with the United States, will prove an additional link between the nations whose friendship is founded upon their common interest and reciprocal esteem."

First Message over the Atlantic Cable—August 16, 1858.

Dedication.

———o———

The publishers of this work have great satisfaction in being permitted to dedicate this volume to the man whom the public recognise as the real author of the Atlantic Telegraph—

CYRUS WEST FIELD.

New York, *August*, 1858.

CONTENTS.

APPENDIX.

ILLUSTRATIONS.

THE STORY OF THE TELEGRAPH.

CHAPTER I.

THE SCIENCE OF TELEGRAPHY—ITS INCEPTION AND PROGRESS—GRADUAL DEVELOPMENT AND PERFECTION.

THE completion of the Atlantic Telegraph, the unapproachable triumph which has just been achieved in the extension of the submarine electrical Cable between Europe and America, has been the cause of the most exultant burst of popular enthusiasm that any event in modern times has ever elicited. So universal and joyful an expression of public sympathy betokens a profound emotion that will not immediately pass away. The laying of the Telegraph Cable is regarded, and most justly, as the greatest event in the present century; and it is with the desire to meet the popular demand for an authentic and concise history of this great event that the authors of this volume have undertaken their task, and not with the expectation that they shall be able, in the very brief time afforded them, to present the world with a volume entirely worthy of the importance of the subject. The history, such as it is, will at least have the merit of correctness.

The completion of the Atlantic Telegraph may be regarded as the crown and complement of all past inventions and efforts in the science of Telegraphy; for great and startling as all past achievements had been, so long as the stormy Atlantic bade defiance to human ingenuity, and kept Europe and America dissevered, the electric Telegraph was deprived of the crowning glory which its inventor had prophesied it should one day possess. But now the great work is complete, and the whole earth will be belted with the electric current, palpitating with human thoughts and emotions. If we reflect for a moment that the great Atlantic Cable is the connecting link between America's web-work of forty-five thousand miles, and Europe's system of fifty-five thousand miles of Telegraph wires, thus forming a vast inter-connected system of a hundred thousand miles of wires, more than sufficient to put a quadruple girdle round the globe, some conception of its immense significance may be gained.

In this history, it is the aim of the authors to include

within the scope of their work an account of the development of the Telegraphic system, its beginnings and applications, its rapid improvements and almost miraculous extension over the civilized parts of the earth.

Of all the marvellous achievements of modern science, the Electric Telegraph is transcendently the greatest and most serviceable to mankind. It is a perpetual miracle, which no familiarity can render commonplace. This character it deserves from the nature of the agent employed and the end subserved. For what is the end to be accomplished, but the most spiritual ever possible? Not the modification or transportation of matter, but the transmission of thought. To effect this an agent is employed so subtle in its nature, that it may more properly be called a spiritual than a material force. The mighty power of electricity, sleeping latent in all forms of matter, in the earth, the air, the water; permeating every part and particle of the universe, carrying creation in its arms, it is yet invisible and too subtle to be analysed. Of the natural effects of electricity, the most palpable examples occur in atmospheric manifestations; but its artificial generation and application are the mightiest scientific triumphs of our epoch. It was but little more than a hundred years ago that Franklin's immature experiments demonstrated the absolute identity of lightning and electricity. Since then various mechanical contrivances have been devised for liberating this subtle but potent power from its dark windings in the prison-house of material forms; the result of which is, that the electric fluid may be produced and employed in any desired quantity and with any required intensity. Thus the same terrific agent which rushes with blinding and crushing force in the lightning, has been brought under the perfect control of man, and is employed at his will as an agent of his necessities. With dissolving energy it effects the most subtle chemical analyses, it converts the sunbeam into the limner's pencil, employs its titanic force in blasting rocks, dissolves gold and silver, and employs them in the gilding and plating of other metals; it turns policeman, sounding its whistle and alarm-bell; and lastly, applies its marvellous energy to the transmission of thought from continent to continent with such rapidity as to forestall the flight of Time, and inaugurate new realizations of human powers and possibilities.

The efficacy of the Electric Telegraph depends on the power to produce at will the three following effects:—

1st. To develope the electric fluid in any desired quantity.

2nd. To transmit it to any required distance without any injurious diminution of its force.

3rd. To cause it upon its arrival at any required point to produce some sensible effects which may serve the purpose of written or printed characters.*

* Lardner The Electric Telegraph.

Every practical application must have its ground and genesis in some scientific conception; it must pre-exist in the mind as law, before it can assume substantive shape in the world of concrete realities. Thus practical navigation is the result of mathematical discoveries and observations, that run back to the speculative labors of the Greek geometers; and our ships now navigate the trackless ocean with safety, guided by a knowledge of the principles of conic sections discovered by Apollonius and Aristarchus. A practical embodiment is real and lasting, just in proportion to its truthful relation to laws generalized from the observation of phenomena in nature, and any discovery is explained, when the ideal steps on which it depends, are set forth in systematic order.

It was not until the latter part of the eighteenth century, that the science of electrology began to receive some of those great generalizations which give it a rational character, and which, in fact, constitute it a science. The first serviceable steps were the distinction of the two electricities, MUSCHENBROEK'S experiments with the Leyden Jar, and FRANKLIN'S great meteorological discovery, which was the first manifestation of the influence of electricity in the general system of nature. These were followed up by the vast labors of COULOMB and AMPÈRE, bringing electrical phenomena under the jurisdiction of mathematics. In the year 1820, OERSTED published to the world his beautiful and comprehensive discovery, connecting the laws of Electricity and Magnetism. Ten years afterwards, ARAGO and FARADAY came with their brilliant intuitions, bringing those grand generalizations which have been the foundation of all the magnificent applications of the science which have since been made.

Such is a brief and rapid view of the development of the science of Electrology. How practical applications have kept pace with abstract conceptions, and the energy and enterprise of intelligent men have been all the while fully abreast with the discoveries of science, remains to be proved.

It would seem to be necessary to the perfection of every great discovery, that it should pass through a series of rudimentary and embryonic stages before it can gain a serviceable and rational form. Through such stages did the applications of steam pass, as witness the numerous experiments for centuries previous to its receiving the foundation in science, from which alone we derive all our power over this force. Telegraphy, too, has had to pass through analogous processes of development. To the present generation, who have seen this greatest of modern arts grow up under their own eyes within the past ten or twelve years, it can hardly seem possible that they have been present at the very birth and adoption of this great idea. But, notwithstanding that the art is so new, and has been so suddenly brought to perfection, the idea is old, and, like other great ideas, has had to struggle through long ages for its perfect development. There were many abortive experiments through the century and a half preceding the first practical success. Fruitless though the greater part of these experiments were, yet they were all necessary or inevitable to the final triumph. And as this History will be chiefly occupied with the triumphs of the telegraphic art during the past twelve years, under the guidance of the great scientific laws previously evolved, it will be necessary to take a glance at the preliminary endeavors towards the consummation of the great idea; though from the imperfect development of the science of Electrology, large and permanent results were not possible.

THE History of Telegraphy may properly be divided into three periods:

1st. From the development of electricity by friction to the discovery of Galvanism, or the production of Electricity by the chemical union of acids upon metals, in 1790 by Galvani, and by Volta in 1800.

2d. From the discovery of the Galvanic or Voltaic Battery at the beginning of the present century, including the discovery of Electro-Magnetism by Oersted in 1820, and Ampère's first application of the principles he evolved, up to 1831, when Professor Henry discovered the method of constructing improved magnets in connexion with properly arranged batteries, so as to produce mechanical effects at a distance.

3d. The Era of application, from 1837, when Professor Morse in America, and Cook and Wheatstone in England, respectively patented their telegraphic inventions, and inaugurated the triumphant and almost miraculous successes which the past twelve years have witnessed.

In the year 1726 JOHN WOOD, of England, discovered that electricity could be conveyed a long distance by conducting wires, and in 1747 one of the earliest applications of the discovery was made by Doctor WATSON, who extended his experiments over a space of four miles, comprising a circuit of two miles of wire and an equal distance of ground.

In 1784* M. Lomond, of France, communicated telegraphic signals to a neighboring room by means of a pith-ball electronometer, acted upon by electricity, an account of which is narrated in "Young's Travels in France." And, according to the *Comptes Rendus*, *Séance* 1838, M. Belancourt in 1798 established a telegraph between Madrid and Aranjuez, twenty-six miles in length, through which a current of electricity was forced and gave signals for letters.

* Phil. Transactions, Vol. XIV.

The first Galvanic Telegraph of which we have any account was constructed by Söemering, of Munich: it operated by the decomposition of water, and caused a bell at the opposite end of the wire to ring. This was the first decomposing or chemical telegraph, and it can even now be operated, according to "Jones's Book of the Telegraph," though less rapidly than Bain's.

The year 1820 was signalized by the discovery of electro-magnetism by Professor Oersted, of Copenhagen. This most important discovery was at once seized upon by M. Ampère, and embodied in the first Electro-Magnetic Telegraph. This, however, proved more an experimental than a practical advance in the science.*

The next advance was made by Mr. Sturgeon, of England, who constructed the first electro-magnet by rolling a piece of copper wire around an iron of a horseshoe form. He found that when the electric fluid passed through the coil the inclosed iron became a magnet, and was again demagnetized in breaking the current. Additional advances were made in 1831, by Professor Henry, who discovered a method to which we have already alluded, of forming magnets of great intensity, making practicable the production of powerful effects at a great distance. This was indispensable to the creation of electro-magnetic telegraphing for great distances, and

* Annales de Chemie et de Physique, 1820.

was, of course, a *sine quâ non* to the possibility of that crowning achievement of science, the Submarine Telegraph.*

In the year 1823, Gauss and Weber first constructed the simplified Electro-Magnetic Telegraph. It was Gauss who first employed the incitement of induction, and who demonstrated that the appropriate combination of a limited number of signs is all that is required for the transmission of messages. Weber discovered that a copper wire, 7,400 feet long, which he carried over the houses and church steeples of Göttingen, from the Observatory to the Cabinet of Natural Philosophy, required no special insulation. This was a most important discovery in the construction of telegraphic lines, and has been of immense service in the art of Telegraphy.

Such were some of the preparatory steps through which the telegraphic art passed previous to the inauguration of the great era commencing in 1837. Thus we see that the mighty achievements of the past twelve years were the results of the conspiring labors and investigations of many generations of patient workers, who were denied the gratification of witnessing the final glories of their discoveries.

The world has now more than a hundred thousand miles of Electric Telegraph. To say that this achievement marks an era in social life, is not to give it the

* Silliman's Jour. Vol. XIX.

proper characterization. It marks an area in the unfolding of the human mind. The Telegraph has more than a mechanical meaning; it has an ideal, a religious, and a prospective significance, far-reaching and incalculable in its influences.

The inspired author of the Book of Job exclaims in an interrogatory, meant to bear the burden of the impossible, "Canst thou send lightnings that they may go, and say unto Thee, here we are?" But this is precisely what science has done in the Electric Telegraph. In all our cities there are buildings in the cellars of which machinery exists for the fabrication of lightning, which is supplied to order, at a very moderate price, in any quantity required, and of any desired force, which is conducted for thousands of miles across rivers, through forests, over mountains, and down through the dark depths of the ocean. And this lightning is made the vehicle of thought, to carry messages to the extreme ends of the earth, between two beats of the pendulum of a clock. The fabled horses of Arabian tales, and the famous legend of le Beau Pecopin's midnight ride round the world, are tame in the comparison of the realities of Telegraphy.

It has been the result of the great discoveries of the past century, to effect a revolution in political and social life, by establishing a more intimate connexion between nation and nation, with race and race. It has been found that the old system of exclusion and insulation, are stagnation and death. National health can only be maintained by the free and unobstructed interchange of each with all. How potent a power, then, is the telegraphic destined to become in the civilization of the world! This binds together by a vital cord all the nations of the earth. It is impossible that old prejudices and hostilities should longer exist, while such an instrument has been created for an exchange of thought between all the nations of the earth.

Such is the vista which this new triumph of the might of human intelligence opens to us. Every one must feel stronger and freer at the accession of such an increase of power to the human family, as has been conferred upon it by the success of the Ocean Telegraph. It shows that nothing is impossible to man, while he keeps within the sublimely imperious orbit of Nature's laws.

> " The future hides in it
> Gladness and sorrow:
> We press still thorow,
> Naught that abides in it
> Daunting us, Onward."

CHAPTER II.

LAND AND OCEANIC TELEGRAPHS.

THE entire history of the Magnetic Telegraph is compressible within very narrow limits. The first Telegraphic line in the United States was erected only fourteen years ago. But twenty-one years have passed since the first English patent for a Telegraph was issued. A period of thirty-nine years has elapsed since the discovery and first application of electro-magnetism. A space of a trifle over a third of a century, therefore, embraces the era of Telegraphic operations. The accomplishment of the last great feat of underlaying the ocean suggests the propriety of a retrospect of early attempts.

The discovery of electro-magnetism is due to Professor OERSTED, of Copenhagen, who announced the new principle in 1819. The discovery was seized by M. AMPERE, the eminent French physicist, who in the following year, invented an electro-magnetic telegraph, in which he used as many wires as there were letters, and broke and restored the circuit by keys, similar to those used in the House patent. This attempt was purely experimental. It was never practically tested. No current was obtained of sufficient force to traverse any considerable distance:—the idea of using the earth to complete the circuit; the possibility of employing a single wire; any method of recording the magnetic current, in other words, of not only making it speak, but of reporting and preserving its utterances, all these were unknown elements, which it was left for the present generation to discover. The first advance was made by Professor JOSEPH HENRY, then of Princeton College, now of the Smithsonian Institution, who, by the construction and novel combination of magnets, in the year 1831, demonstrated the possibility of transmitting the current over long distances; a revelation indispensable to the construction of a submarine telegraph. In 1833, WEBER, a German experimenter, found that a copper wire which he carried over sundry houses and church steeples of Gottingen, required no especial insulation; a fact of great practical value to telegraphing upon land. The year 1837 furnished several additions to previous knowledge; and, in fact, may be regarded as the epoch of the inland telegraphic system. In July of that year, STEINHEIL put in use a registering electro-magnetic telegraph between Munich and Bogenhausen, wherein clock-work was employed to pass a ribbon of paper through the machine under a deflected needle, which impressed upon it dots and marks, accepted as representations of the several letters of the alphabet. A few days before the Steinheil apparatus was set to work, Messrs. COOKE and WHEATSTONE obtained their English patent for a telegraph using a deflective point, the patent bearing date, June 12, 1837. Their specific improvement was the use of transmitting or relay magnets.

In the year 1835, Mr. SAMUEL F. B. MORSE, of New York, constructed a rude apparatus for telegraphic experiments in the University of the City of New York. Seventeen hundred feet of wire were stretched around the walls of a small apartment, and connected with a recording machine of rough construction. This experiment proved the practicability of the Telegraph. The first word indicated through the action of the electric current was "Eureka." Mr. MORSE conducted further experiments until the year 1837, and in October of that year filed his caveat for the "American Electro-Magnetic Telegraph," in which an incomplete outline of his actual system was presented. He represented that his plan had been devised in the year 1832, but was then first reduced to the test of experiment. Dr. CHARLES T. JACKSON, of Boston, has always contended that the MORSE invention was due to his suggestion, made to the Professor during a voyage from Europe to the United States, on board the ship *Sally*, in the Summer of 1832.

There is no proof, however, to contradict the averments of both gentlemen, that they had heard nothing of the Steinheil and Wheatstone inventions. MORSE obtained his patent in France, in 1838, and in 1840 a patent in the United States. In 1846, a re-issue of the latter patent was obtained, in which the claim to the electric or magnetic current was abandoned, but he claimed instead the use of electro-magnetism as a motor. The same year he patented a right to the invention of a local circuit. Subsequently, Mr. ALEXANDER BAIN patented, in England, his claim for an improved Electro-Chemical Telegraph, where the message was recorded by electricity upon paper chemically prepared; and in 1848, entered his claim for an American patent, which was confirmed in 1849. In 1848–9, Mr. ROYAL E. HOUSE, of New York, obtained an American patent for a Telegraph, in which the message was recorded by types, and the circuit broken and resumed by means of keys similar to those of the piano-forte, answering to the letters of the alphabet.

The first electro-magnetic line in the United States was that between Baltimore and Washington, the distance forty miles, completed in 1844. Congress contributed $30,000 towards its construction. The first message over this line was sent by Miss ANNE ELLSWORTH, of Connecticut, on the 27th May, 1844, and the words transmitted were these four: "*What hath God wrought?*"

The operation of this initial enterprise promising success, a company was formed, with Mr. AMOS KENDALL as President, for the continuation of the line; and in 1845 it was extended between New York and Wilmington, Del., leaving a gap between the latter point and Baltimore, which was filled up early in 1846. From this inception, the work has advanced until the present day, when there are more than thirty-five thousand miles of telegraph lines in the United States, connecting the coast of Newfoundland with the shores of Texas, and the great plains of the West, and the great lakes of the North with the Atlantic and the Gulf of Mexico. There are more than five thousand miles in the British Provinces; in England there are over ten thousand miles; and in the world a total length exceeding one hundred thousand miles.

The lines of Telegraph now in operation in the United States, are (1) Morse's; (2) Bain's; (3) House's; (4) Hughes'. The latter is a new invention, possessing wonderful sensitiveness, and combining the advantages of Morse's and House's. A general description of these different systems may be usefully introduced in this connexion.

The engraving exhibits the Register of the Morse Telegraph, as used in the telegraph offices:

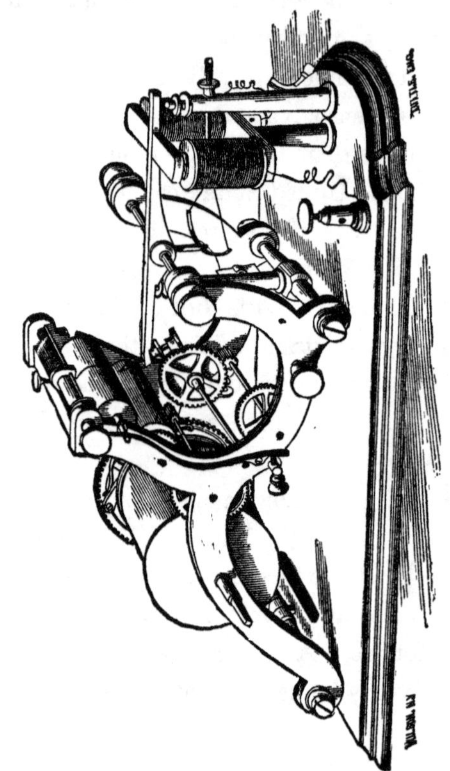

REGISTER OF THE MORSE TELEGRAPH.

In this illustration, the magnet, the armature, the rollers, and the clock-work, are shown. The machine is set in operation by a lever-key, placed at the other end of the telegraphic route, which, being raised or lowered by the pressure of a finger, breaks or closes the circuit. A signal-key is also used, and the apparatus for recording messages is simple and effective. The subjoined illustrations convey an idea of these parts of the machine:

SIGNAL-KEY OF MORSE'S INSTRUMENT.

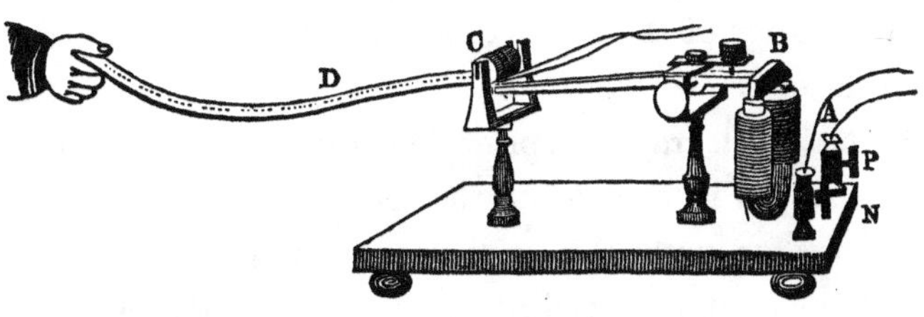

RECORDING APPARATUS.

The writing by Morse's instrument is a series of dots and dashes, a full description of which may be found in the Appendix.

Bain's Telegraph is a modification of Morse's. Its form is shown in the following cut:

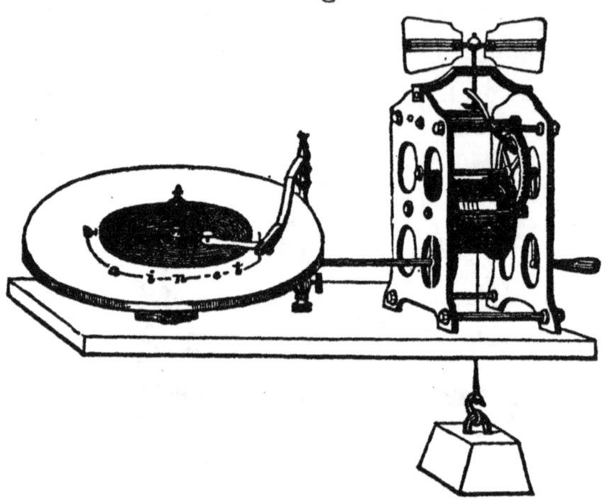

BAIN'S TELEGRAPH.

In this Telegraph, chemically prepared paper is marked by the passage of the current, and the message is recorded upon the disc.

House's Telegraph is a printing instrument. Its general character is shown in the subjoined engraving. The operator with this instrument manipulates a lettered key-board, arranged like a piano-forte; the circuit being closed by pressing down the keys; a type-wheel revolving at the extremity of the line, records the message in printed Roman letters.

Hughes' Telegraph resembles House's, and, like that, prints its messages. The principal advantage claimed for this instrument, is its peculiar delicacy; a feebler current of electricity sufficing to set it in motion. In principle, it is a combination of the Morse and House Telegraphs.

HOUSE'S TELEGRAPH.

The method of erecting a line of Land Telegraph is so familiar, that any description is superfluous. The operation of splicing the wires, at points of junction, is not, however, so generally known. It is exhibited in the accompanying engraving.

Submarine Telegraphs have a very recent history.

One of the earliest difficulties to be overcome in terrestrial telegraphing, was the extension and perfect insulation of the wire over streams and sheets of water. At first, the transit was effected by using bridges, where bridges existed; and in their absence, of suspending the wires over the water, from carefully-selected prominences on either bank. In time, the non-conducting quality of

SPLICE OF THE WIRES IN A LAND TELEGRAPH.

water suggested the idea of submerging the line, and permitting it to sink to the bed of the stream; and with the application of india rubber or gutta percha, as a coating to prevent oxidation, the plan was successfully resorted to.

The Cable generally used for river crossings has the following size and shape:—

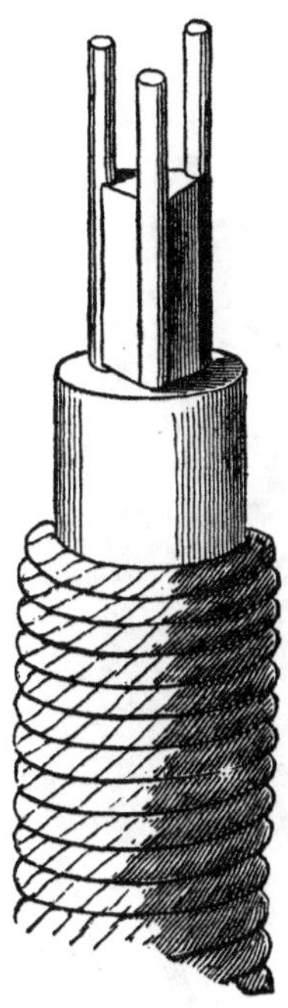

CABLE FOR RIVER CROSSINGS.

The employment of Submarine Cables for telegraphic communications was first successfully accomplished seven years ago. In October, 1851, a deep-sea Cable was laid in the English Channel, between Dover and Calais. This Cable had four conducting wires, insulated by gutta percha, and afterwards enveloped by tarred rope-yarn and galvanized iron wires. Its general plan of construction is indicated in the engraving:

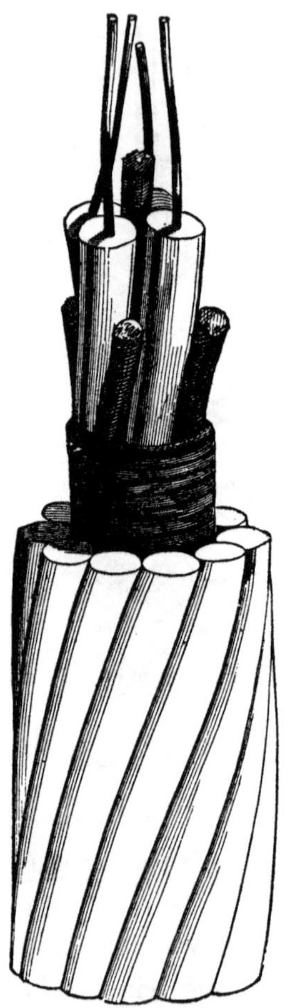

SUBMARINE TELEGRAPH CABLE, CONNECTING DOVER AND CALAIS.
EXACT SIZE.

This Cable was manufactured in the space of three weeks. It weighed seven tons to the mile, and was twenty-four miles in length. It will be observed that the principle differs essentially from that of the Atlantic Cable; four conducting wires being used instead of seven, and the aggregate weight being six times greater. Owing, however, to the chafing of the wire upon the rocks near the French coast, this Cable severed at the end of a month, and a new and stronger Cable had to be laid. This is now in perfect working order.

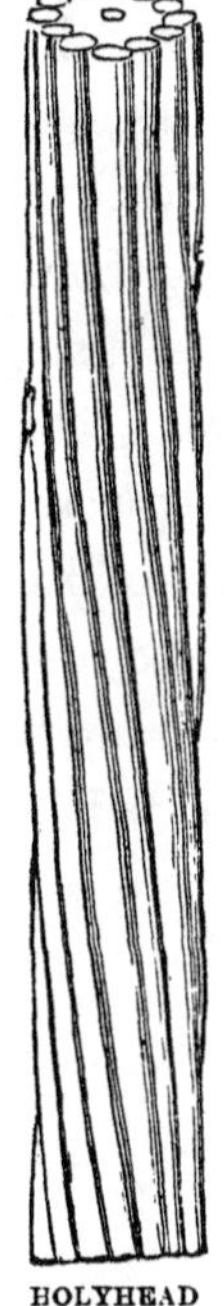

HOLYHEAD SUBMARINE CABLE.

A similar Cable was soon after made and laid down by R. S. NEWALL & Co., between Holyhead and Dublin, which worked perfectly for several days; after which its insulation became imperfect. Its size and form are exhibited in the accompanying cut.

A Cable entirely of hemp, without any galvanized wire covering, was laid down between Portpatrick and Donaghadee by the same firm, for the Magneto-Electric Telegraph Company. This has entirely failed.

Including the Atlantic Cable, the aggregate length of the Submarine Telegraph lines of the world, is now little short of three thousand miles.*

* Appendix—"Table of Submarine Telegraphs."

The immediate result of the first apparently successful attempt with the Cable across the Straits of Dover, was the suggestion of various projects of a similar character. The plan of a trans-Atlantic Cable does not seem to have been among these. The idea was too stupendous, perhaps, and seemingly impracticable to be conceived; or, if conceived, to be entertained otherwise than as a desirable impossibility. In 1851, however, a speculator was found bold enough to propound the enterprise, using the columns of the London *Athenæum* for the purpose. He proposed to use a single stout wire, enveloped, firstly, in a gutta-percha coat, and secondly in hemp, saturated with some imperishable matter, and to extend it directly from the coast of Ireland to Newfoundland. The suggestion fell still-born,—only, however, to be revived in a year or two afterwards, under the auspices of the Company of whose history it is now time to treat.

CHAPTER III.

ORIGIN OF THE ATLANTIC TELEGRAPH—ORGANIZATION OF THE NEW YORK, NEWFOUNDLAND, AND LONDON TELEGRAPH COMPANY.

CONFLICTING claims are always set up for the honors justly due to the originators of useful enterprises. Crude ideas, when first broached, rarely receive the degree of attention to which they are often really entitled, and it is not unfrequently the case that the actual projector of a plan of vast magnitude finds an incredulous audience to receive his demonstrations. In the history of the inception of the Atlantic Telegraph, it is probable that many new elements will enter. The credit of the original invention of Submarine telegraphing will undoubtedly be claimed by various parties. Had this wonderful work proved a total failure, aspiring inventors would perhaps have been less anxious to claim its paternity. Having become a fact in the history of the world, it is not a matter of surprise to find a host of rival claimants springing up; each pressing his demand for priority, and each unwilling to yield to the pretensions of others. We do not propose to enter into any elaborate discussion of this knotty question. The great fact remains unaltered, that a Submarine Oceanic Telegraph is not only possible, but actual. It is idle to attempt to compress within the compass of a single chapter any complete record of the conflicting claims which are put forward in connexion with the story of this undertaking; indeed, a work much larger than the present one would scarcely suffice for the presentation of the plans for which their authors require the endorsement of the public. We, therefore, content ourselves with a general summary of the facts of the case, which seem, after careful comparison of data, and conscientious investigation of the merits of opposing claims, to be established beyond the reach of cavil.

The question of the priority of discovery of the principle of the Electro-Magnetic Telegraph, as lying between Prof. MORSE, Prof. HENRY, and Dr. JACKSON, does not properly enter into this department of the history of Telegraphing. The merits of the claims set up for these parties are treated elsewhere. For the present, we have to deal solely with the record of the origin of Submarine Telegraphs; and in order to arrive at a satisfactory conclusion in regard to this particular branch of the subject, it is essential to refer briefly to events which occurred at intervals from the years 1847 to 1856, a period covering some nine years. While disclaiming any intention to slight the claims of ingenious inventors, whose skill and industry will insure them the grateful remembrance of posterity, even if their names be disconnected from the historical record of the Atlantic Telegraph, we are led to the belief that the credit of the inception, progress, and successful completion of that great undertaking, which forms the existing link between Europe and America, is due to the foresight, prudence, and unwearying energy of three or four gentlemen, all of whom have contributed to the enterprise the results of long experience and the fruits of enlarged scientific knowledge.

One fact should be stated at the outset. It is undoubtedly true that the success of Submarine Telegraphing depends upon a single point. That point, once gained, insures other conditions, necessarily consequent upon it. In other words, no submarine cable for telegraphic purposes can be perfect until its insulation is rendered positive. But one material is known to possess this insulating property. But for the discovery of gutta percha, the Atlantic Telegraph would not have been worked; the electric current would have been dissipated in the depths of the sea; the triumph of mechanical skill and scientific genius, over which two nations have become ecstatic, could not have been accomplished.

Prior experiments, on shorter lengths of submarine cables, demonstrated the useful properties of this new material. From these early attempts sprang the project for underlaying the ocean. Diligent industry, the application of fertile resources, and the hearty co-operation of two countries in the work, have made the Atlantic Telegraph the fitting climax to a long series of careful investigations. The utility of the insulating material, known as gutta percha,* has been abundantly tested, both by scientific experiment and in practical service. But a few years have elapsed since its introduction as an article of trade; fewer still have passed since its suitability as an insulating material for telegraphic wires was first definitely established. The credit of the discovery of

* "*Gutta Percha.*—A valuable substance, known only within the last few years. It is the concrete juice of a large tree (*Isonandra gutta*), growing in certain parts of the Malayan Archipelago. The first specimen of the inspissated juice which appeared in England, was presented to the Society of Arts in 1843, but two or three years elapsed before a just sense of the importance of the substance began to gain ground. In 1845 the importation of gutta percha into England amounted to only 20,600 lbs.; in 1848, it had reached 3,000,000 lbs.; in 1851, it amounted to 30,580,480 lbs. The honor of having drawn attention to its real nature and uses is due to Drs. D'Almeida and W. Montgomerie. The purposes to which gutta percha is applied, are too numerous for recapitulation. It resists the action of water, and is at the same time a bad conductor of electricity; it is, therefore, employed for enclosing the metallic wires used in the Electric Telegraph. The efficiency of the Submarine Telegraph is largely due to this valuable substance."—*Homans' Cyclopædia of Commerce.*

this peculiar virtue seems to be justly awarded to Mr. S. T. ARMSTRONG, of the City of New York. This gentleman was invited to visit England in the year 1847, for the purpose of examining the new material, then just coming into notice as an article of commerce. The practicability of its application to many useful purposes was considered settled. A new branch of trade appeared to be opened by its discovery. A company was formed in New York, of which Mr. ARMSTRONG became President. The first shipment made from England to the United States, was an invoice of five tons, which was received here in the year 1847. Various experiments demonstrated the utility of the new material for manufacturing purposes, but it was not until the autumn of 1848, that the insulating property was so far developed as to be relied upon with certainty. At that period, a number of experiments were made, the result of which proved that copper wires became perfect conductors of electricity when coated with gutta percha, resisting the action not only of the air, but of the water; and that a telegraphic wire, encased in this material, became a safe conductor of an electric current under conditions which would otherwise prove an insuperable bar to success. This was the germ of the Submarine Telegraph, and it would be unjust to Mr. ARMSTRONG to detract from the merit to which his early investigations fairly entitle him.

Next came the practical solution of the problem. In

this branch of the subject the first practical experimenter seems to have been a telegraphic agent in an office at Montreal, Mr. F. N. GISBORNE. Other persons had conceived general ideas of the principles of constructing oceanic telegraphs; but the facts in the history of early experiments upon this point demonstrate that the first practical application of the principle, at least on this side of the Atlantic, was made by Mr. GISBORNE. In 1851–2, Mr. GISBORNE, then recently from England, went to Halifax, and thence to New Brunswick and the United States, endeavoring to find responsible parties who would undertake the work of laying a submarine line. He was unsuccessful in this quest; but in a few months afterwards received partial aid, and accomplished the experiment of laying a small insulated Cable from the main land to Prince Edward Island. His next step was to lay a submarine line from Newfoundland to Cape Breton, and in a preliminary survey he underwent many hardships. In the interval which elapsed before arrangements could be made for perfecting this project, his backers failed. In the Spring of 1854 Mr. GISBORNE came to New York, placed himself in communication with Mr. CYRUS W. FIELD, enlisted the sympathies of other influential gentlemen, and finally received an appointment as Superintendent of the Company which was formed about that time to establish a line of Telegraph between Nova Scotia and Newfoundland.

The connexion of Mr. CYRUS W. FIELD with the Atlantic Telegraph enterprise, therefore, dates from the early part of the year 1854. Receiving with undoubting faith the plan for connecting the continents by means of an Oceanic Telegraph, seeing no obstacles which could not be overcome by patient perseverance, and possessed of an indefatigable energy, to Mr. FIELD may be accorded the honor of sustaining the main burden of an extraordinary effort. When others sank, discouraged by the pressure of untoward events, and dismayed by the prospect of failure, this gentleman revived hopes that were nearly extinguished, infused fresh energy into the efforts of his associates, and finally succeeded in arousing a spirit of enterprise which has reaped its own reward. The history of the organization of the Telegraph Company, and the record of the steps in the progress of the Atlantic Telegraph are so intimately associated with the name of Mr. FIELD, that we may be pardoned for a brief digression from the main subject of this narrative, in order to give a running sketch of that gentleman's personal history.

CYRUS WEST FIELD is a native of Massachusetts, having been born in the town of Stockbridge, in that State, in the year 1819. His father was the Reverend D. D. FIELD, a native of East Guilford, Connecticut, a graduate of Yale, and first settled at Haddam, Ct. Dr. FIELD had nine children—seven sons and two daughters. The sons have all risen to distinguished positions. The elder brother, the Hon. DAVID DUDLEY FIELD of New York, is well known on both sides of the Atlantic as one of the Revisers of the Code of the State of New York. MATTHEW DICKINSON FIELD is a leading citizen of Massachusetts, and was recently or still Senator. JONATHAN EDWARDS FIELD is a Judge of the Supreme Court of California. The Rev. HENRY M. FIELD was formerly Pastor of a Congregational society in West Springfield, Massachusetts, and now Editor of the *New York Evangelist.* One son, TIMOTHY, went to sea, many years since, and has never been heard from. CYRUS WEST FIELD, in early life, came to New York, and was engaged as clerk in the establishment of Mr. A. T. STEWART. He subsequently returned to Massachusetts, and was employed in the paper manufactory of his brother MATTHEW, in the town of Lee; and on attaining his majority entered into the same line of business on his own account, at Westfield, Massachusetts, but failed during the panic of 1837. He then returned to New York, and established a large paper commission warehouse, of which he is still the head. Some four or five years ago, Mr. FIELD'S attention was directed to the project of an Oceanic Telegraph. In the spring of 1854, his ideas on that subject first took definite shape, and the active and earnest coöperation of several prominent citizens of New York—among whom were Messrs. PETER COOPER, MOSES TAYLOR, MAR-

SHALL O. ROBERTS, CHANDLER WHITE, S. F. B. MORSE, and DAVID DUDLEY FIELD—was given in aid of his enterprise. The further development of the plan is recorded in these pages.

In person, Mr. FIELD is slight and nervous. His weight is about one hundred and forty pounds. His features are sharp and prominent, the most striking peculiarity being the nose, which projects boldly. His body is lithe and his manner active; eyes grayish-blue and small; forehead large, and hair auburn and luxuriant. He does not appear as old as he is. The steel portrait which accompanies this volume conveys a perfect idea of the appearance of the man.

Another name,—that of Professor MORSE,—has been intimately associated with the early history of the Atlantic Telegraph, and merits particular mention. Although not actively connected with the last stages of that undertaking, yet Professor MORSE has freely given his co-operation and sympathy to it; while the acknowledged value of his services in the cause of Telegraphy entitles him to grateful remembrance. SAMUEL FINDLAY BREESE MORSE, like Mr. FIELD, is a native of Massachusetts. He was born at Charlestown, Mass., on the 29th April, 1791; graduated at Yale College in 1810; and then went to London to study the art of painting under BENJAMIN WEST. Returning to the United States in 1815, he began the practice of his art in the city of New York, and about the year 1820 was one of the founders of the National Academy of Design. He revisited Europe in 1829, and on his return to America in 1832, seems to have worked out the plan of an Electro-Magnetic Telegraph; the honor of which invention, however, is claimed by Dr. JACKSON. Of this point, we treat briefly elsewhere. Since the year 1835, the attention of Prof. MORSE has been chiefly directed to Telegraphic operations; and during the past year a handsome remuneration has been voted him by the Continental Governments.

We return to the narrative of the primary stages of the Telegraphic enterprise.

The organization of the New York, Newfoundland, and London Telegraph Company dates back to the year 1854. In March of that year, Mr. CYRUS W. FIELD, his brother, DAVID DUDLEY, and Mr. CHANDLER WHITE were commissioned to proceed to Newfoundland, to obtain from the Government of the Province an act of incorporation. On arriving at St. John's, they called upon the Governor, who convoked the Executive Council the same day. The Governor gave a favorable answer to the Commissioners, and immediately sent a special message to the Legislature, then in session, recommending them to pass an act of incorporation, with a guaranty of interest on the Company's bonds to the amount of £50,000, and a grant of fifty square miles of land on the island of

Newfoundland, to be selected by the Company. These terms were agreed upon.

Additional grants were subsequently received from the Governments of Prince Edward Island, Nova Scotia, Canada, and the State of Maine; and afterwards from the Governments of Great Britain and the United States. The results of these negotiations may be summarily indicated, for future reference, in this place, as upon the liberal nature of the grants depended the ultimate results of the project. The governmental grants extended to the Company, from first to last, have therefore been as follows:—

NEWFOUNDLAND.

Exclusive privileges for fifty years of landing Cables on Newfoundland, Labrador, and their dependencies.

The exclusive right embraces a coast line extending from the entrance of Hudson's Straits southwardly and westwardly along the coasts of Labrador, Newfoundland, Prince Edward Island, Cape Breton, Nova Scotia, and the State of Maine, and their respective dependencies.

Grant of fifty square miles of land on completion of Telegraph to Cape Breton.

Similar concession of additional fifty square miles when the Cable shall have been laid between Ireland and Newfoundland.

Guarantee of interest for twenty years at five per cent. on £50,000.

Grant of £5,000 in money towards building road along the line of the Telegraph.

Remission of duties on importation of all wires and materials for the use of the Company.

PRINCE EDWARD ISLAND.

Exclusive privilege for fifty years of landing Cables.

Free grant of one thousand acres of land.

A grant of £300 currency per annum for ten years.

CANADA.

Act authorizing the building of telegraph lines throughout the Provinces.

Remission of duties on all wires and materials imported for the use of the Company.

NOVA SCOTIA.

Grant of exclusive privilege for twenty-five years of landing Telegraphic Cables from Europe on the shores of this Province.

STATE OF MAINE.

Similar grant of exclusive privilege for like period of twenty-five years.

GREAT BRITAIN.

Annual subsidy of £14,000 sterling until the net profits of the Company reach 6 per cent. per annum, on the whole capital of £350,000 sterling, the grant to be then reduced to £10,000 sterling per annum, for a period of twenty-five years.

The aid of two of the largest steamships in the English navy to lay the Cable, with two subsidiary steamers.

A Government steamship to take any further necessary soundings, and verify those already taken.

UNITED STATES.

Annual subsidy of $70,000 until the net profits yield 6 per cent. per annum, then to be reduced to $50,000 per annum, for a period of twenty-five years, subject to termination of contract by Congress after ten years, on giving one year's notice.

The United States steamship *Arctic* to make and verify soundings.

Steamships *Niagara* and *Susquehanna* to assist in laying the Cable.

A Government steamer to make further soundings on the coast of Newfoundland.

The original organization of the Company was as follows:

NEW YORK, NEWFOUNDLAND, AND LONDON
TELEGRAPH COMPANY.

DIRECTORS IN NEW YORK:

Peter Cooper, Cyrus W. Field,
Moses Taylor, Marshal O. Roberts,
Chandler White.

Peter Cooper,	President.
S. F. B. Morse,	Vice President.
Moses Taylor,	Treasurer.
Chandler White,	Secretary.
David Dudley Field,	Counsel.
F. N. Gisborne,	Engineer.

The first step in the great enterprise, now fairly inaugurated, was the connexion of St. John's with the

Telegraphic lines already in operation in Canada and the United States. The first attempt to lay these wires was made in 1855, but it then proved unsuccessful. In 1856 the effort was renewed with success, and there has been little interruption of the union between the two islands. The Cable employed for this service is quite large, composed of three strands, and has three conducting wires. From Port-au-Basque, the Cable station on the western part of Newfoundland, the telegraph extends directly across the island to Trinity Bay, the American terminus of the Atlantic Telegraph.

In the year 1856, the Company dispatched Mr. CYRUS W. FIELD to England to enlist the aid of capitalists in that country. The most complete success attended his efforts. The capital stock of the New York Company was fixed at $1,750,000, and the whole was subscribed for,—one hundred and one shares in London, eighty-eight in America, eighty six in Liverpool, thirty-seven in Glasgow, twenty-eight in Manchester, and the remainder in other parts of England. The capital, however, had to be subsequently increased to $2,500,000, to meet the failures that occurred in the various attempts to submerge the Cable.

The project, when brought to the notice of the British and American governments, was received with a like degree of favor, and liberal subsidies were granted; the substance of which has already been indicated.

'he Act of Congress, approved March 3, 1857, and Charter of Incorporation, granted by Parliament, are en in the Appendix. The stipulations contained in se acts form an interesting part of the general history he Telegraph.

'he Charter of the New York, Newfoundland, and idon Company, conferring upon it the exclusive right and telegraphic cables on the shores of Newfoundland other parts of North America, and for twenty-five rs to do the same thing on the shores of Nova Scotia, made over to the "Atlantic Telegraph" Company—Direction of which is now constituted as follows:

Chairman.
SAMUEL GURNEY, M.P., London.

Vice-Chairman.
T. H. BROOKING, London.

Directors.

BRETT, J. W., London.
BROWN, WILLIAM, M.P., Liverpool.
DUGDALE, JAMES, Manchester.
HANKEY, T. A., London.
HARRISON, HENRY, Aigburth, near Liverpool.
HORNBY, THOMAS DYSON, Liverpool.
JOHNSTON, EDWARD, Liverpool.
LAMPSON, C. M., London.
LE BRETON, FRANCIS, London.
LOGIE, WILLIAM, Glasgow.
PEABODY, GEORGE, London.
PENDER, JOHN, Manchester.

PICKERING, C. W. H., Liverpool.
SCHWABE, GUSTAV CHRIS., Liverpool.
THOMSON, Professor W., LL.D., Glasgow.
ARCHIBALD, HON. E. M., H.M. Consul, New York.
BELMONT, AUGUSTE, Banker, New York.
COOPER, PETER, Merchant, New York.
CORBIN, FRANCIS P., New York.
HUNT, WILSON G., Merchant, New York.
LOW, A. A., Merchant, New York.
MORGAN, MATTHEW, Banker, New York.
SHERMAN, WATTS, Banker, New York.
CARTIER, HON. GEORGE E., Quebec, Lower Canada.
ROSS, HON. JOHN, Toronto, Upper Canada.
YOUNG, HON. JOHN, Montreal, Upper Canada.
ROBERTSON, HON. JOHN, St. John, New Brunswick.

General Manager: CYRUS W. FIELD.*
Engineer: CHARLES T. BRIGHT.
Electrician: E. O. W. WHITEHOUSE.
Secretary: GEORGE SAWARD.
Solicitors: FRESHFIELDS & NEWMAN.
Auditors:—JONATHAN RIGG, No. 17 Mark Lane, London, Merchant; HENRY W. BLACKBURN, Bradford, Yorkshire, Public Accountant.
Bankers: THE BANK OF ENGLAND.

The New York Company also made over to the new Corporation all concessions bearing upon the undertaking which may be hereafter obtained, and all the patent rights of Messrs. WHITEHOUSE and BRIGHT,

* Resigned.

which in any way concerned the working of instruments in marine circuits of great length, were prospectively secured to it. In order that the capital subscribed might be entirely applied to the immediate object of the undertaking, the projectors, Messrs. BRETT and FIELD, and Messrs. BRIGHT and WHITEHOUSE, considerately arranged that compensation for the privileges assigned, and for past expenditure and exertions, should be left entirely dependent on the successful result of the undertaking. The final agreement with these gentlemen was, that upon attaining success, a half-yearly dividend of ten per cent. per annum on the capital should first be paid to the shareholders, and then one-half of any further profit should be given to them, and the other half be retained by the Company, it having been estimated upon a very moderate computation of the probable amount of revenue, conjoined with a consideration of the comparatively small working expenses, where there can only be two terminal stations to be maintained, that a very satisfactory result might be secured to all parties upon this ground.

But while the electrical and financial preparations had terminated so favorably to the views of the Company, there were other topics of equal moment not yet satisfactorily determined. The solution of one momentous question remained to be given. Could a telegraphic wire be laid on the bottom of the Atlantic? Every care was, therefore, taken to bring together all the evidence that could be gleaned of the actual character of the vast oceanic basin, which was to be the scene of the great enterprise, and to collate them with the labors of Lieutenant MATTHEW F. MAURY, who had already demonstrated the existence of an *Atlantic plateau.**

This plain, according to Lieut. MAURY, was scarcely twelve thousand feet below the level of the sea, and extended in a continuous ledge from Cape Race, in Newfoundland, to Cape Clear, in Ireland. Its greatest depression was declared to be in mid-ocean, whence it imperceptibly ascended to the shore on either side. In order to verify the theory of such a plateau, the aid of the government of the United States was invoked by the Company. A cordial assent was given; and Lieut. O. H. BERRYMAN, U.S.N., was twice dispatched in the steamer *Arctic* to make soundings along the proposed line; while, to verify his observations, Her Britannic Majesty's steamer *Cyclops* traversed the ground in an opposite direction. The knowledge thus obtained was conclusive. The plain was gently levelled, so deep as to be below the reach of disturbing superficial causes, and composed of particles of shells, so minutely triturated as to render their character indetectible save with the aid of a microscope. Their presence, examined by the lights of science, proved how little those profound depths had been disturbed in the course of uncounted

* See Appendix.

ages, and encouraged the hope that the Cable, when once laid along with them, might rest as tranquilly—perhaps as long. The tendency of these infinitesimal fragments to agglutinate to any metallic centre exposed to them, held out the expectation that the submerged Cable would soon be thickly enveloped by them, and a fresh element of security so obtained. The accompanying map comprises a complete view of the plateau, as it stretches from shore to shore.

This submarine plateau is really a gently-levelled plain, lying just so deep as to be inaccessible to the anchors of ships, and to other sources of surface-interference, and yet not so far depressed but that it can be reached by mechanical ingenuity without any very extravagant effort. It seems, indeed, that it is a portion of a great zone of table land, which entirely engirdles the earth, or which at least stretches from the western side of America to the Asiatic coasts of the Pacific.

CHAPTER IV.

THE ATLANTIC CABLE—CONSTRUCTION AND EXPERIMENTS.

IN the construction of the Atlantic Cable, many important considerations were necessarily taken into account. There were certain characteristics which the Cable must possess, to enable it to meet the peculiar circumstances of the case, and the conditions in which it would be placed. The success of any plan for the laying of an Oceanic Telegraph was believed to be greatly dependent upon the form and character finally given to the Cable itself. Before the terms and details of the contract could be satisfactorily presented to contractors, it was essential to compare different plans of construction, and decide upon that which promised the most effective results. The Directors of the Company gave patient attention to the proposals which were laid before them, and after a careful examination of the relative merits of plans submitted for their adoption, awarded the contract for the construction of the great Submarine Cable to the firm of GLASS & ELLIOT, of Greenwich, near London. The beautiful workmanship of this Cable is not less creditable to the establishment in which it was manufactured, than honorable to the scientific skill and assiduity of Mr. GLASS, the senior partner of the firm, to whom the Directors unanimously accorded the praise due to his indefatigable exertions in their interest. A contract for the construction of one-half of the Cable was subsequently awarded to Messrs. R. S. NEWALL & Co., of Birkenhead.

The general plan of the Cable having been adopted, certain specific calculations became necessary. The first important point to be settled was the weight of the Cable. While it must be sufficiently heavy to sink quickly to the bottom of the sea by its gravity, when launched from the stern of the paying-out vessel, it was requisite that any excessive weight should be avoided; else the difficulty of management in the deep sea would become an obstacle almost insuperable. The Directors, in announcing to the stockholders the results of their long investigation, dwelt with much earnestness upon the difficulty which they encountered, in the commencement of the enterprise, in the determination of this delicate problem. They cited the account given by Mr. BRETT, of his unsuccessful attempt to connect Europe with Africa by a Cable of massive construction; and argued from the experience of that gentleman, that the management of heavy Cables in the ocean would be an impracticable undertaking. If, on the contrary, the Cable were too light, it would be at the mercy of the currents, its integrity would be greatly risked, its strands might be separated, and its insulation destroyed. Again, it was obviously desirable that, size and specific weight being given, the Cable should be made as strong as material and dimensions allowed. Its positive requirements were tenacity and flexibility. The ingenious combination of these qualities with a perfect electrical condition, which were attained as the result of the careful experiments of Mr. GLASS, aided by distinguished scientific gentlemen, justified the choice of his plan by the Directors of the Company.

The Atlantic Cable, now lying at the bottom of the ocean, is an extremely simple contrivance. No alteration has been made in its construction during the entire progress of the Telegraph Expeditions. Severe tests have failed to develope defects in its practical operation; electrical experiments have established its fitness for the purpose designed; the frigate *Niagara* has tested its strength by swinging to it as though at anchor in mid-ocean; its wonderful flexibility has been proved by repeated trials. Had the Atlantic Telegraph enterprise developed only this remarkable result of mechanical ingenuity, the work would not have been undertaken in vain. A slender thread, laid by powerful mechanism at the bottom of a vast ocean, and laid without a flaw or break, linking two worlds together in bonds of amity, and marking a new era in the history of the earth, is in itself a triumph.

The illustration on page 62 shows the exact size of the Atlantic Cable.

The profile view of the Cable (p. 63) gives a general idea of its appearance when ready for use. In order to show more fully the process of manufacture, an illustration of sections of the Cable is given on page 63.

The central conducting wire is a strand made up of seven wires of the purest copper, known in the trade as No. 22. The strand itself is about the sixteenth of an inch in diameter, and is formed of one straightly drawn wire, with six others twisted round it; the twisting having been accomplished by dragging the central wire from a drum through a hole in a horizontal table; the table itself revolving rapidly, under the impulse of steam, carrying near its circumference six reels or drums, each armed with copper wire. Each drum revolved upon its own horizontal axis, and delivered its wire as it turned. This twisted form of the conducting wire was first used in the Submarine Cable laid across the St. Lawrence in 1856. It was then employed with a view to the reduction to the lowest possible amount of the chance of an interruption of continuity. It was considered improbable that a fracture would occur in more than one of the wires in this twisted strand at precisely the same spot; so that, although the whole seven wires might be broken at different parts of the strand, the capacity of the Cable for the transmission of the electric current would not be destroyed. During the process of manufacture at Greenwich, the copper used in the construction of the Atlantic Cable was assayed from time to time in order to insure absolute homogeneity and purity. Experiments upon the strand itself proved that, when subjected to strain, it was capable of stretching 20 per cent. of its length without breakage, and without material interference with its conducting power.

This yielding temper in a strand of pure copper inspired grave doubts in the minds of many gentlemen connected with the early stages of the undertaking. It was anticipated that when the Cable was subjected to strain, the yielding core would become attenuated to such an extent that its capacity for the transmission of a current would be virtually destroyed. To meet this objection, and dispel the growing apprehension, Mr. WHITEHOUSE, a capable electrician, who had taken an active part in the scientific investigations pertinent to this undertaking, devised a simple and very effective experiment. He connected three lengths of the Cable of 200 miles each into a continuous line, and then passed a current from two 36-inch double induction coils excited by 10 Smee cells, each having plates of 100 square inches of area, through the 600 miles of Cable to the magneto-electrometer. The weight of 745 grains was raised on the end of the steel yard, and was thus the measure of the current after transmission through the Cable. He next made a break in the Cable at the distance of 400 miles from the nearer end, and introduced into the gap one mile of fine insulated wire, which possessed only one-eleventh of the capacity of the copper strand. This proportion was ascertained by weighing equal lengths of the wire and the strand. The piece of wire weighed three grains, and the piece of strand weighed thirty-three and a half grains. A current from the same induction coils was now again passed through 600 miles length of Cable to the magneto-electrometer, with the one-mile length of fine wire interpolated in its course, and 725 grains were lifted on the steel-yard. Only twenty grains of lifting power out of a force equivalent to 745 grains had been lost in consequence of the introduction of the mile of fine wire, measuring but one-eleventh of the central strand. The fear that a stretch of two feet in a mile for six miles of the Cable would render it electrically unfit for service, was thus met by showing that, if the entire copper strands of the Cable were stretched 96 feet in every mile, the loss of conducting capability would amount to no more than a thirty-seventh part.

A subsequent experiment determined the fact that the copper strand bore twenty per cent. of the elongation without injury to its integrity of texture, or in other words, it could be stretched one thousand feet in a mile not only without breaking, but without impairing its telegraphic utility. The copper strand, indeed, was never broken until elongated to the extent of twenty-five or thirty per cent. These experiments having satisfied the incredulous—a troublesome class of persons who always swarm upon the track of a new invention, and whose little faith is sometimes a serious bar to progress—the construction of the Cable was pushed forward with remarkable vigor. The general plan of manufacture is exhibited in another page. The following is a vertical section of the Atlantic Cable, showing the position of the central conducting wires, with their coverings of gutta-percha, rope-yarn, and twisted wires.

VERTICAL SECTION OF THE ATLANTIC CABLE.—EXACT SIZE.

The principal processes through which the Cable passed were four in number—1, the twisting of the conducting wires; 2, a triple coating of gutta-percha; 3, a covering of fine thread yarn soaked in a mixture of pitch, tar, oil, and tallow; 4, the final enclosure of twisted wire.

We shall describe these processes in their order. The copper strand of the Cable having been prepared in the manner already indicated, was rolled upon drums as it

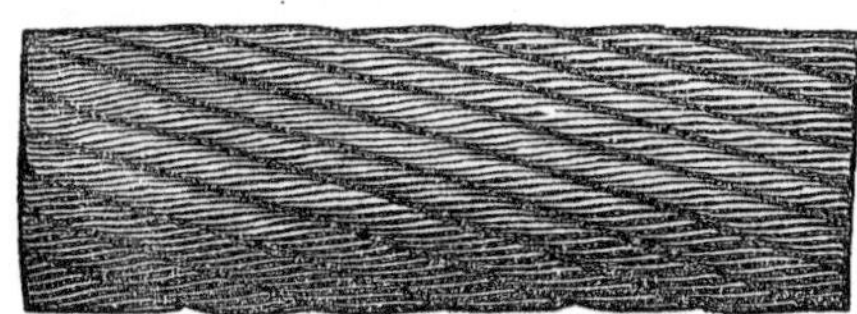

PROFILE VIEW OF THE ATLANTIC CABLE—EXACT SIZE.

VIEW OF THE ATLANTIC CABLE IN SECTIONS.

1. Exterior covering of wires, eighteen in number, of seven strands each.
2. Covering of tarred rope-yarn.
3. Three coatings of gutta-percha.
4. Copper conducting wires, seven in number.

was completed, in lengths of two miles. It was taken from these drums to receive a coating of three separate layers of refined gutta-percha. The original diameter of the conducting wire before this coating was one-sixteenth of an inch. After receiving the coating, the diameter was increased to three-eighths of an inch. These preliminary processes were by far the most important of the whole, for the perfection of the insulation of the Cable depends upon the integrity of the insulating material. Three coatings of gutta-percha were applied at suitable intervals to insure the efficiency of the work. The gutta-percha employed for the purpose was prepared with the utmost possible care. Lumps of the crude substance were first rasped down by a revolving toothed cylinder placed within a hollow case. The raspings were then passed between rollers, and macerated in hot water; afterwards washed in cold water, and driven, at a boiling water temperature, by hydraulic power, through wire-gauze sieves, attached to the bottom of wide vertical pipes. The gutta-percha came out from these sieves in plastic masses of remarkable purity and fineness. It then passed into an apparatus known as a masticator, consisting of a series of interrupted screws revolving in hollow cylinders; the material being squeezed and kneaded for some hours in this manner, in order to expel the water and render the substance perfectly homogeneous. Horizontal cylinders heated by steam received the purified gutta-percha. Screw-pistons driven down slowly, but with resistless force, pressed the material through a die, which at the same time had the strand of copper wire moving along through its centre. The strands entered the die naked, bright copper wire, and emerged as thick, dull-looking cords, having received one complete coating. The same process was repeated, until three coatings inclosed the copper strands.

The Cable, having been prepared thus far in lengths of two miles, rigorous tests of insulation and electric continuity were applied. Each length was coiled on a wooden drum, with a short piece of the copper conductor projecting at each end. These drums were then immersed in water, and the task of the Electricians began. The continuity was ascertained by passing a voltaic current of low power through the strand, from a battery of a single pair of plates, and causing it to record a signal after issuing from the wire. The amount of insulation was determined by a different plan. One pole of a voltaic battery, consisting of 500 pairs of plates, was connected with the earth; the other pole was united to a wire coiled around the needle of a sensitive horizontal galvanometer, and running thence to the exposed strand of the Cable, which was left without any conducting communication. If the insulation was perfect, the earth formed one pole of the battery, and the end of the insulated strand the other pole, the circuit remaining open: consequently no current passed, and the needle of the galvanometer was not deflected in the slightest degree. If the insulation was imperfect, or there was undue electrical permeability in the sheath of gutta-percha, a portion of the current forced its way from the strand through the faulty places in the covering of gutta-percha, and the needle of the galvanometer was deflected; the degree of deflection being the measure of the amount of imperfection. It was found that the best coating was not a thorough insulation, a slight deflection being produced in the needle, but insufficient to cause serious interference with telegraphic operations. A certain degree of deflection, therefore, was considered allowable and safe. It was only when this degree was exceeded that the core was condemned. While the test for continuity was absolute, that which determined the insulation was in a measure relative. A very powerful battery was used in the tests for insulation, in order to render the trial as severe as possible. During the progress of these experiments, an ingenious method was adopted for the purpose of testing at the same time both the continuity and the insulation. The operation was as follows: The entire length of the Cable was joined into a loop or endless ring, when a voltaic sand-battery of 500 pairs of plates was connected by one of its poles with the entirely insulated strand of the Cable, and by its other pole with the earth. The circuit was thus insulated as a whole, and charged as a Leyden jar. But a charged Leyden jar may be made a part of a voltaic circuit; and therefore this charged ring of wire was able to transmit a low-tension circuit without its charge being interfered with. A small insulated battery was then introduced into the circuit, and its low current flowed from pole to pole through the strand.

A bell, also insulated, was so placed in the same circuit that any break of continuity dropped a needle previously held by magnetic attraction, released some wheel-work, and sounded an alarm; the bell was consequently heard whenever the continuity of the strand failed. Another bell was so placed as to be rung whenever the current from the five-hundred-cell battery acquired undue power in consequence of faulty insulation.

Electrical experiments having finally established the perfection of continuity and insulation, the Cable was now ready to undergo the process of joining the lengths. The two-mile coils of completed and proved core were wound on large drums, with projecting flanges on each side, the rims of which were shod with iron tires, so that they could be rolled about as broad wheels. When the core was in position on these channelled drums, the circumference of each drum was closed in carefully by a sheet of gutta-percha. The work of the gutta-percha manufacturers ended with this final preparation. The core-filled drums passed from their hands into the custody of the joiners. Each drum was then mounted with axles, the gutta-percha covering removed, and the projecting ends of the copper strands carefully brazed together. This process may be described as follows: A piece of copper wire was attached by firm brazing an inch or two beyond the point of junction on one side, tightly wound round until it reached to the same extent on the other side, and was then firmly brazed on again. A second piece of copper wire was then brazed over the first in the same fashion, and extended a little way beyond it; and finally several layers of gutta-percha were carefully laid over and around the joint by the use of hot irons. This operation is identical with that of splicing the Cable, which has been repeatedly effected with entire success, and by means of which the laying of the wire in mid-ocean was accomplished during the last voyage of the *Niagara* and *Agamemnon*. A clear idea of the stages of this delicate manipulation is given in the subjoined illustration.

The explanation of this method of splicing the Cable, which has already been given, will suffice for a comprehensive view of a part of the Telegraphic enterprise upon which depended the success of the whole. It will be seen by reference to the cut, that the electrical connexion must be preserved even if the joint in the Cable yields. In the event of a rupture of the Cable, by which the core on each side should be dragged opposite ways, the electric condition would still remain perfect. The outer investment of the wire would unroll spirally as the ends

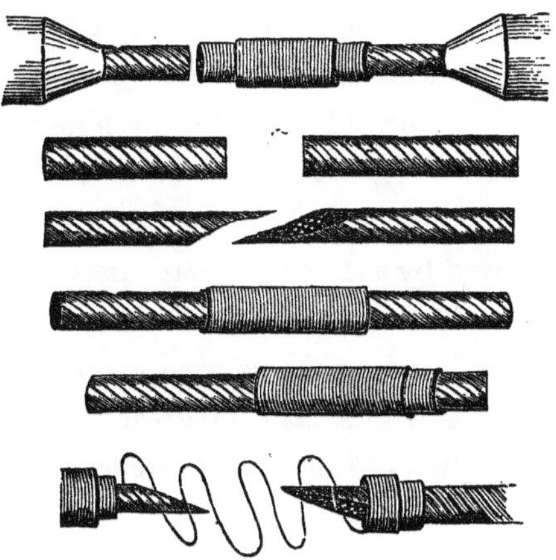

THE SPLICE OF THE CABLE.

of the Cable were pulled asunder; so that however the mechanical continuity of the strand itself might be broken, the conducting power would still remain.

After the lengths had been joined in the manner indicated, the Cable underwent another process, passing to a "serving" machine, fitted with a horizontal wheel, on which were placed five bobbins. Each bobbin was supplied with some hundreds of yards of five-thread rope-yarn, prepared for the purpose by a previous immersion in a mixture of pitch, tar, oil, and tallow. The coated wire moved slowly through the centre of the wheel, and as it passed up, the bobbins, revolving at the rate of three hundred and seventy-five times a minute, spun the five strands of yarn tightly about it, not leaving the smallest interstice. At this stage of preparation, the Cable passed from this machine through a gauge, which showed its diameter to be nine-sixteenths of an inch; while the electric current with which it was in connexion, proved by the needle of the galvanometer, that the connexion and insulation of each fathom as it moved off was uninjured by the serving process.

The Cable being now in a state of great forwardness, it only remained to "close" or bind it up in wire. For this purpose another horizontal table, arranged like the one for the serving process, was provided. It carried near its circumference eighteen bobbins or drums; each drum filled with bright charcoal-iron wire, and having two motions, one round its horizontal axis, and one round an upright pivot, inserted into the revolving table, so that the strand was delivered always towards the centre of the table as it was carried swiftly round. The iron strand was of the same diameter as that which was used for the copper core, each strand consisting of seven iron wires. With each whirl of the closing-machine, therefore, eighteen iron strands were firmly twisted round the central core. The core, actuated by the rollers of the machinery, rose through the middle of the table and ascended towards the ceiling; the metallic twist, as it passed, tightly embracing it. One hundred and twenty-six iron wires were, therefore, woven about the Cable in order to complete the process of its manufacture. Twenty-one of these machines were kept constantly at work in the factory of Messrs. GLASS & ELLIOTT, and about an equal number in NEWALL'S establishment at Birkenhead. The labor accomplished at GLASS & ELLIOTT'S establishment in the course of twenty-four hours, by the use of twenty-one machines, was as follows: Two thousand and fifty-eight miles of iron wire were daily twisted into two hundred and ninety-four miles of strand; this length of strand sufficing to cover about ten miles of the Cable.

The Cable thus completed was drawn from the closing-machines at the rate of thirty feet per minnte, or eighteen hundred feet per hour, passing through a gauge of five-eighths of an inch diameter. It was then carried by revolving wheels through a tank of hot tar, issuing forth into the yard thoroughly fitted for the duty to which it was to be devoted. In the yard it was coiled away in huge piles, ready for shipment; each day's labor adding some inches to the height and bulk of the mass.

The weight of the main Cable was eighteen hundred and sixty pounds, or nearly one ton, to the mile. For the shore-ends, a heavier style was provided, of augmented dimensions and greatly increased power of resistance, but constructed upon the same general plan. The weight of the shore-ends now in use, is seven tons to the mile, and the diameter at the shore is about one and a half inches, tapering through about half a mile into the main Cable. The following engraving exhibits a vertical section of the shore-end, of the exact size. This part of the Atlantic line is encased by twelve solid charcoal-iron wires of No. 1 gauge. The No. 1 wires pass gradually into No. 2, and No. 2 into No. 3, as deep water is reached. The gutta-percha casing and serving of tarred hemp are also thicker upon these "shore-ends" as well as the outer iron coat.

VERTICAL SECTION OF SHORE END OF ATLANTIC CABLE—EXACT SIZE.

The Directors of the Company, in their official manifesto, published last year, took occasion to explain the reasons why a return circuit had not been provided in planning this Cable. It was well known that in every land telegraph yet brought into use, the earth itself had been found sufficient for the completion of the circuit, and hence a return wire could not be deemed absolutely essential. Moreover, the addition of a second wire would have added largely to the size, weight, and expense of the Cable, and would have unavoidably deferred the completion of an Atlantic Telegraph to another year.

The total cost of the Atlantic Cable was nearly five hundred dollars per mile. The aggregate outlay of the Company in the year 1857, on account of the construc tion of the Cable alone, was stated as follows:—

Price deep-sea wire per mile, ...	$200
Price spun-yarn and iron wire per mile, . . .	265
Price outside tar per mile	20
Total per mile, ...	$485
For 2,500 miles, ...	$1,212,500
For 10 miles deep-sea Cable, at $1,450 per mile,	14,500
For 25 miles shore-ends, at $1,250 per mile, .	31,250
Total cost	$1,258,250

The scientific experiments which were undertaken by competent electricians in the employ of the Company established the fact, that a wire across the Atlantic was not only practicable, so far as mechanical possibility was concerned, but that the scientific difficulties, which were naturally suggested as the effect of distance, and the peculiar conditions in the sea, were not insuperable.

A general view of the results of careful experiments,

which finally decided the Directors upon the adoption of the plan of an Atlantic Cable, now successfully carried out, presents a record of industrious scientific application that may not inappropriately form a part of the history of the enterprise.

In the ordinary arrangement of the wires of the Electric Telegraph, where they are stretched upon posts, and insulated by glass and the surrounding air, the current of electricity runs along as a simple stream, and with a velocity that is almost inappreciable for ordinary distances. But when the wires are inclosed in a sheath of insulating substance, like gutta-percha, and placed in a moist medium, or a metallic envelope, the case is very different. The influence of induction then comes into play as a retarding power. As soon as the insulated central wire is electrically excited, that excitement operates upon the adjoining layer of metal or moisture, and calls up in it an electrical force of an opposite kind. Each of these forces disguises, or holds fast, an equivalent portion of the other,—and the electricity of the central wire is thus prevented from moving freely onward, as it otherwise would. It is found, in short, that the Submarine Telegraph Cable is virtually a lengthened Leyden jar, and transmits signals while being charged and discharged, instead of merely by allowing a stream of the electrical influence to flow dynamically and evenly along it. And every time it is used it has first to be filled and then emptied. In the case of a long submarine wire, this was found to be a task requiring considerable time,—and this was found, moreover, to be very much increased with an increase in the length of the wire. And when experiments came to be made in 1851, upon telegraphic lines running underground, between London, Manchester, and Glasgow, and upon others partly underground, and partly submarine, between London, Paris, and Brussels, it was found that the speed of the current did not exceed 1,000 miles per second. In that year, Mr. WHITEHOUSE invented a very ingenious method of determining with precision the force of currents thus transmitted; and the result of his investigations was to show, that in submarine telegraphs the wires act as reservoirs, and not as mere channels,—that the larger reservoir receives and holds a larger quantity than the smaller one, and that this naturally produces the most powerful effects when allowed to escape from its imprisonment. By 1855, the scientific corps provided themselves with much more complete and perfect instruments for pursuing these inquiries, and the construction of new Telegraph lines also furnished them with better opportunities of making their experiments. It was soon found that a magneto-electrical current took a second and a half to discharge itself, when it moved through 1,146 miles of wire, in consequence of the retarding power of induction in this extended medium. This was a rate of speed not at all compatible with any profitable employment of a Transatlantic Telegraph for commercial purposes,—and the next step was to devise some remedy for this inductive obstacle. The first thing done was to send different kinds of electricity along the wire in succession, in the hope that each transmission of one kind would clear away the residue of the other which had immediately preceded it. The result was a complete success. Although the same wire, and the same magneto-electric combination were employed, which had before demanded a second and a half for the completion of a single discharge, seven and eight currents now readily recorded themselves in a single second. When positive followed negative, and negative followed positive, in exactly equal proportions, the electrical equilibrium of the wire was continually restored as fast as it was disturbed—each current clearing away the inductive influence which the other had left behind it. It was proved, moreover, in the course of these experiments, that successive charges of electrical influence,—either of the same kind, or of alternate opposite kinds,—may be travelling along lengthened conducting wires simultaneously, the one following the other, like successive waves upon the sea. Alternate positive and negative signals were sent along 900 miles of wire, at the rate of eight signals in each second,—and two signals arrived at the end of the wire after the acts of transmission had been discon tinued. In another experiment, by the use of a wire,

1,020 miles long, three signals of a single-stroke bell were distinctly heard after the movement of the hand which originated the current had ceased. This, therefore, indicated a way in which the rapidity of transmitting electrical currents along a submarine wire could be increased; it was necessary only to employ opposite kinds, —positive and negative alternately.

The next point to be investigated was the ratio in which increase of distance in a gutta-percha covered telegraph wire augments the difficulties of rapid transmission. It had been supposed that the available force was diminished in the ratio of the square of the distance traversed,—that is, that a current which has traversed 600 miles has only a thirty-sixth part of the working force of a precisely similar current which has travelled only 100 miles. In experimenting upon this point they had to consider: First—the diminution of the current's power to produce mechanical effects; and, Second—its loss of speed. A voltaic battery of 72 pairs of plates, each with a surface of 16 inches, was set to work, and it was ascertained how many grains the current would raise upon being transmitted through a wire just long enough to effect the connexion. The number of grains lifted was 25,000. The experiment being repeated with the same current through 200 miles of wire, the number of grains lifted was 10,650; with 400 miles of wire it was 3,250; and with 600 miles it was 1,400. Clearly the loss of mechanical power in this case was not diminished in so large a ratio as had been supposed. In regard to loss of speed, nearly five thousand observations were made, with wires varying in length from 83 to 1,020 miles, to determine its ratio; and from these it appeared that with a wire 83 miles long the transmission was effected in .08 of a second; with 166 miles in .14 of a second; with 249 miles in .36 of a second; with 498 miles in .79 of a second; and with 1,020 miles in 1.42 of a second. This ratio was far less than had been supposed. The result of the experiments was to establish, with tolerable accuracy, the fact that the velocity of movement of a magneto-electric current, through a gutta-percha covered wire, is 300 miles in from one-twelfth to one-sixteenth of a second; 600 miles in from one-sixth to one-ninth of a second; and 900 miles in from one-fifth to one-fourth of a second. Still further experiments proved that a rate of transmission could be obtained by the employment of magneto-electric currents from two and a half to three times as great as that of any simple voltaic impulse which can be used. The maximum speed attained by voltaic electricity was 1,800 miles per second; the maximum for the magneto-electric current was 6,000 miles per second. This showed conclusively that this is the current which must be employed. And it was also established that large coated wires, used beneath the water or the earth, are worse conductors, so far as velocity of transmission is concerned, than small ones; and it was this which led to the adoption of the small-sized copper wire which was finally decided on as the conductor by the Atlantic Telegraph Company.

After these points had been established by experiment —rendering it theoretically probable that there would be no difficulty in using a wire, if it could once be laid down across the Atlantic—the next point was actually to record a signal by a current sent through a circuit of 2,000 miles. For this purpose, in 1856, the various lines of Telegraph under charge of the English and Irish Magnetic Telegraph Company were used, and they are so extensive, have so many ramifications, and each line contains so many separate wires, that a continuous length of nearly 5,000 miles could be made up among them. The experiments were made with great care, under the supervision of Mr. BRIGHT, the Engineer; and Mr. WHITEHOUSE, subsequently the Electrician of the Company. On the 9th of October, 1856, ten gutta-percha covered wires, each measuring over 200 miles, were connected, so that a continuous circuit was formed of above 2,000 miles, and signals were distinctly and satisfactorily telegraphed through the whole length, at the rate of 210, 241, and, upon one occasion, 270 per minute. Experiment having shown that the conditions present in insulated wires placed under the ground and beneath the sea are strictly analogous, this result was regarded as establishing, beyond all reasonable doubt, the practicability of working the Transatlantic Telegraph.

The Company was therefore indebted to Mr. WHITEHOUSE and Mr. BRIGHT for a series of experiments which established certain important facts. General results may be indicated in a few words—viz:

That gutta-percha covered submarine wires do not transmit as simple insulated conductors, but that they have to be charged as Leydèn jars, before they can transmit at all.

That, consequently, such wires transmit with a velocity that is in no way accordant to the movement of the electrical current in an unembarrassed way along simple conductors.

That magneto-electric currents travel more quickly along such wires than simple voltaic currents.

That magneto-electric currents travel more quickly when in high energy than when in low, although voltaic currents of large intensity do not travel more quickly than voltaic currents of small intensity.

That the velocity of the transmission of signals along insulated submerged wires can be enormously increased from the rate, indeed, of one in two seconds, to the rate of eight in a single second, by making each alternate signal with a current of different quality, positive following negative, and negative following positive.

That the diminution of the velocity of the transmission

of the magneto-electric current in induction-embarrassed coated wires, is not in the inverse ratio of the squares of the distance traversed, but much more nearly in the ratio of simple arithmetical progression.

That several distinct waves of electricity may be travelling along different parts of a long wire simultaneously, and within certain limits, without interference.

That large coated wires used beneath the water or the earth are worse conductors, so far as velocity of transmission is concerned, than small ones, and therefore are not so well suited as small ones for the purposes of submarine transmission of telegraphic signals; and

That by the use of comparatively small coated wires, and of electro-magnetic induction-coils for the exciting agents, telegraphic signals can be transmitted through two thousand miles with a speed amply sufficient for all commercial and economical purposes.

About the time that the manufacture of the Cable was completed, the *London Times* rather startled its readers by the announcement that the enterprise must necessarily prove a failure. "It will scarcely be credited," said that journal, "but it is nevertheless true, that the twist of the spiral wires of the Birkenhead half of the Cable is in exactly the opposite direction to the twist of the wires made at Greenwich. Thus, when joined in the centre of the Atlantic, they will form a right and a left-hand screw, and the tendency of each will be to assist each other to untwist and expose the core. By attaching a solid weight to the centre joining, it is hoped this difficulty and danger may be overcome; but none attempt to conceal that the mistake is much to be regretted. We are informed that Messrs. GLASS and ELLIOTT had nearly one hundred miles of their portion completed before Messrs. NEWALL commenced theirs, and that, therefore, the fault rests with those who began theirs last."

This attack called forth a reply from Messrs. NEWALL, who denied the conclusions at which the writer of the article in question had arrived, holding that the so-called "blunder" was literally of no importance. At the same time they exculpated themselves from blame on the ground that they were acting throughout under the direct instructions and supervision of the engineer of the Company, and that the fault, if there was any, was his.

It may not be uninteresting to give a general description of the machinery contrived for paying out the Cable during the progress of the first Expedition, in the summer of 1857. But a small space need be occupied in this description, for that piece of machinery, when put to trial, proved so totally inefficient that it was rejected in the following year, and replaced by a new one, a description of which will be found in its appropriate place in the chapter devoted to the Expedition of 1858. The plan of the first machine was briefly as follows:

Four cast-iron sheaves, or cylinders, about five feet in

diameter, were ranged in line with one another, fore and aft. The first, commencing forward, was single-grooved; the second and third were double-grooved, and the fourth was single-grooved. The Cable, as it came up from the hold of the ship, passed over one of the grooves in the second drum,—then under it backwards and over and around the first single drum,—thence it returned over the remaining groove in the second,—then it went directly across to one groove in the third, following but a small arc in its periphery,—thence to the last single drum, and downward around this, back to the preceding double one, and, finally, over the unoccupied groove in that to a *fifth* grooved drum standing out upon rigid arms over the stern, from which it was dropped into the sea. The grooves in all these drums were exactly adapted, in size and form, to the Cable. The passing and repassing of the Cable over them served to afford friction-service for controlling the velocity of the rope in passing out. But additional checks for this purpose were provided. The four drums were so connected by gearing that their motions were exactly coincident—the motion of any one of them involving corresponding motion in all the rest. Upon two of the shafts, moreover, friction-brakes—the same in principle as those used upon railroads—were applied, to control the velocity of the drums; and to these, which were worked by a screw, was attached a balance, which was to indicate the precise amount of strain thrown upon the Cable at any moment. The screw was worked by a crank, at which was stationed an officer whose duty it was to watch the balance and regulate the friction of the brakes accordingly.

The shipment of the Cable speedily followed the completion. The portion received by the *Niagara* was manufactured by Messrs. NEWALL & Co., at Birkenhead, and shipped at that place; and the other half, received on board the *Agamemnon*, was shipped from the works of Messrs. GLASS & ELLIOTT, at Greenwich. The total length of Cable manufactured was twenty-six hundred miles. In order to make room for the immense coils in which the Cable was deposited in the *Niagara*, the fore-hold of that vessel was cleared of the chain-lockers, coal-bunkers, and tanks, and fitted with a level floor over the kelson, the beams having each been trussed with double stays to compensate for the removal of the stanchions. Part of the Cable was also stowed in a space which had been cleared out on the main deck, abaft the engine-room, by displacing some of the officers' berths and encroaching on the ward-room. Three small vessels of 500 tons each were employed to convey the coils from the works of Messrs. NEWALL to the frigate. The part taken on board the *Niagara* was coiled under the superintendence of Mr. WOODHOUSE, C.E., who assisted in laying down the Varna and Balaklava submarine telegraph.

The operation of shipping the Cable was begun in June and completed in the early part of July, 1857. The event was celebrated in England with high festivity and rejoicing. A *fête champêtre* was given on the 23d of July, at Belvidere House, by Sir CULLING EARDLEY; an immense *marquee* pitched upon the lawn in front of the mansion, affording accommodation for some eight hundred and fifty invited guests, among whom were many distinguished gentlemen, both English and American. The unvarying success of the enterprise, thus far, inspired strong hope, and the greetings interchanged on the occasion of this festivity were enthusiastic and cordial to a degree.

In the latter part of July, 1857, the *Niagara* and *Agamemnon* sailed for Queenstown, Ireland, the appointed place of rendezvous. During this voyage, various successful experiments were made. On board the *Agamemnon*, the mechanical appliances for regulating the delivery of the Cable into the sea were kept continually in motion by the small engine on board, which was connected with them, and the whole worked with great precision and facility. The experiments then made by the *Agamemnon* justified hopes of ultimate triumph. A 13-inch shell was attached to the end of a spare coil of the Cable, for the purpose of sinking it rapidly, and was then cast into the sea, drawing after it a sufficient quantity of slack to enable it to take hold of the ground and so set the machinery in motion. The paying out commenced at the rate of two, three, and four knots respectively. The ship was then stopped, and the Cable was hauled up from the bottom of the sea with great facility. The Cable, when recovered, was reported to have been cleaned as bright as the specimens which were distributed among the friends of the enterprise. The exterior coating of tar had been completely rubbed off by being drawn through the sandy bottom of the sea. On the day after this experiment, a length of Cable was run out opposite the Isle of Wight, and was hauled in with perfect ease—the speed in this case having been increased to five knots. On the next day, a mile of Cable was run out and hauled in, while the speed was increased to six and a half knots. The *Agamemnon* then steered for Cork, where her consort, the *Niagara*, had already arrived.

While the *Agamemnon* and *Niagara* lay about a quarter of a mile apart in the Cove of Cork, their Telegraph Cables were passed to each other, and for the first time a circuit was established through twenty-five hundred miles. The *Niagara's* being attached to the galvanometer, and the *Agamemnon's* brought directly to the battery, an electrical current was found to pass immediately, though at first slowly; at once putting at rest the question of transmission through such a length of wire. This demonstration was the more satisfactory, from the fact that the force developed lifted twenty-five grains on Dr. WHITEHOUSE'S galvano-electrometer, when three grains had been found to indicate sufficient power to record intelligible signals. There was no time that night, however, to attach the recording instruments; and when the *Agamemnon* swung at her moorings, she unluckily fouled the wire and broke the connexion. The whole of the next day was spent in recovering and re-uniting the Cable-ends; but, in the meantime, the *Agamemnon* sent aboard a large iron buoy, and several wooden ones, to be used, in case of necessity, for securing the Cable in soundings. On Saturday, August 1, connexion was re-established between the ends, and each of them connected with the earth, as in lines actually laid out. A distinct message was then immediately telegraphed through the whole scope of two thousand five hundred miles—"*Land in sight: all's well*"—were the first memorable words. In this experiment one current occupied, in its passage, an interval of one second and three-quarters; but three successive signals, each perfectly intelligible, could be passed through twenty-five hundred miles in *two* seconds; thus confirming observations made on shorter circuits, by which it appeared that one wire may, at the same instant, be engaged in conveying several distinct electrical waves, with well-marked intervals between them.

It had been at first decided by the Directors of the

Atlantic Telegraph Company, that the *Niagara* and *Agamemnon* should proceed to mid-ocean, and there, having spliced the Cable, separate, and steer, the one for Newfoundland, and the other for the coast of Ireland. At the last hour, however, this plan was altered, though not without some strong opposition in the Board. It was now determined that the *Niagara* should commence laying down the Cable from the Irish coast westward; that she should be accompanied by all the vessels of the fleet, and that upon reaching mid-ocean, the *Agamemnon* should join her Cable to that of the *Niagara*, and complete the connexion by proceeding to the coast of Newfoundland. An argument in favor of this arrangement was, that one end of the Cable being ashore, it could not be *all* lost in event of an accident. It was further contended that by this plan there would be much less weight of Cable to be sustained at any one time. Then the vessels of the fleet would be together, ready to give each other aid in any emergency, and the work, so it was believed, could be more satisfactorily performed by this than by the mid-ocean arrangement.

The scientific arrangements on board both vessels were complete. In the electricians' department, on board of both the frigates, a concerted plan of signals was provided, in order to test the effect of the electric current upon the Cable during every step of the work. These signals indicated time by seconds, and were passed through the whole extent of the wire. At the side of the *Niagara* and *Agamemnon*, patent-logs were placed, which dipped into the sea, and were fitted with vanes and wheels, the latter turning with a degree of velocity exactly proportioned to the rate at which the vessel dragged them through the water. One of these wheels was so arranged as to make and break an electric circuit at every revolution, and record upon the deck of the ship, by apparatus provided for the purpose, the speed of the vessel. A bell also sounded upon every passage of the electric current through the Cable. The brakeman, therefore, watched the balance which indicated the strain upon the Cable, and tightened or relaxed it as occasion required. He was also to listen for the bell, and if at any time its sound ceased—indicating an interruption in the circuit—he was to stop the machinery, the vessel would be backed, and a *winding machine*, provided for the purpose, and worked by a horizontal steam-engine of about 20 horse-power, would be at once set at work, gathering up the slack-rope as the vessel moved astern—the electrician all the while testing the insulating continuity of the Cable, yard by yard, until the defective portion had been discovered. This would then be cut out and the gap supplied by joining up the ends of the uninjured parts, when the paying-out and testing would be resumed as at the first.

Special provision, too, was made for storms. In ordinary weather, or even with brisk strong winds, either ahead or astern, the work could go on without interruption, as the motion would not be so great as to prevent the machinery from retaining complete control of the Cable. But if the wind should blow astern so heavily as to make it necessary for the vessel to come up head to the wind, an apparatus was prepared for paying out over the bow, similar to that already described. And in case a regular gale should arise, strong enough to render it impossible for the vessel safely to retain hold of the Cable at all, preparations were made for abandoning it temporarily. Upon the deck stood two large reels, each wound round with a very strong auxiliary cable, composed of iron wire only, and capable of resisting a strain of ten or twelve tons. Of this there were about two miles and a half on each reel. In case of a heavy storm, rendering necessary the abandonment of the Cable, it would be cut, and the sea end attached to the end of one of these strong iron cords wound upon the reel. This would then be rapidly let out, and the Telegraph Cable lowered to the quiet bottom of the sea, leaving the entire strain of the tempest to be borne by the iron cord. As soon as possible, moreover, the end of this cord would be attached to immense buoys, shaped like the quill-float of the angler's line, and provided with reflectors, so as to be easily seen, which would be tossed overboard, and left to sustain the Cable until the storm should subside, when they would again be picked up, the Cable recovered and rejoined to the part remaining upon the ship, and the work proceed as before.

Such were the preparations and precautions made in the year 1857, for paying out the Atlantic Cable; and complete and perfect as they were then thought to be, yet were they insufficient to insure success.

CHAPTER V.

THE FIRST EXPEDITION—SUMMER OF 1857.

THE first attempt to lay the Atlantic Cable was made early in the month of August, 1857. A period of less than thirty days sufficed for the completion of the final arrangements for this Expedition, the festivities incident to the occasion, the departure of the fleet from Valentia, the trial, the defeat, and the return. At 6 P.M. on Tuesday, August 4, the Telegraph Squadron left Queenstown Harbor for Valentia Bay. It arrived at Valentia on the day following. The fleet detailed for service on this Expedition consisted of eight vessels, four American and four English, as follows:—

1. The U.S. steam-frigate *Niagara*, Captain HUDSON, to lay the half of the Cable from Ireland.

2. The U.S. steam-frigate *Susquehanna*, Captain SANDS, to attend upon the *Niagara*.

3. The U.S. steamer *Arctic*, Captain BERRYMAN, to make further soundings on the coast of Newfoundland.

4. The U.S. steamer *Victoria*, Captain SLUYTER, to assist in landing the Cable at Newfoundland.

5. H.M. steamer *Agamemnon*, Captain NODDAL, to lay the half of the Cable on the American side.

6. H.M. steamer *Leopard*, Captain WAINWRIGHT, to attend upon the *Agamemnon*.

7. H.M. steamer *Cyclops*, Captain DAYMAN, to go ahead of the steamers and keep the course.

8. H.M. steamer *Advice*, Captain RAYMOND, to assist in landing the Cable at Valentia.

The presence on the Island of the representative of Royalty in Ireland contributed in no small degree to the popular idea of the importance of the occasion; and the idea found development in bonfires, pyrotechnic displays, music, feasting, dancing, and cheering, and the characteristic attributes of an Irish merry-making.

His Excellency, the Lord Lieutenant (the Earl of Carlisle), attended by his suite, and accompanied by Sir EDWARD M'DONNELL, Chairman of the Great Southern and Western Railway, several of the Directors of the Company, and Mr. G. E. ILBERY, the courteous and efficient Superintendent of the line, proceeded by special train on Monday morning to Killarney. The Vice-regal party were received at the King's Bridge Station by Mr. ILBERY, and conducted to the state carriage. An elegant *dejeuner* had been provided at Valentia by the Knight of Kerry; the festivities taking place in a large storehouse adjoining the hotel of the place. This storehouse was handsomely decorated for the occasion. It was paved with slabs of slate taken from the extensive quarries in the vicinity, and the tables at which the Company sat were formed of the same material. The banqueting-room was draped with banners, and adorned with evergreens, flowers, and mottoes. At one end, the words of the Irish Welcome, *Cead Mille Failtha*, were prominently displayed; and over the head of the Chairman were placed the flags of the United States and of the United Kingdom, with the initials "J. B." and "V. R." suspended below in handsome wreaths. The Knight of Kerry presided at the banquet, and gave a toast in honor of the Queen, which having been duly honored, the Chairman again rose and proposed the health of the Lord Lieutenant and prosperity to Ireland.

Lord CARLISLE, in responding, made the following eloquent and appropriate remarks:—

"I beg to return you my most hearty thanks for the honor you have done me in so kindly drinking my health. I believe, as your worthy chairman has already hinted, that I am probably the first Lieutenant of Ireland who ever appeared upon this lovely strand. At all events, no Lord Lieutenant could have come amongst you on an occasion like the present. Amidst all the pride and the stirring hopes which cluster around the work of this week, we ought still to remember that we must speak with the modesty of those who begin and not of those who close an experiment; and it behoves us to remember that the pathway to great achievements has frequently to be hewn out amidst risks and difficulties, and that preliminary failure is even the law and condition of the ultimate success. Therefore, whatever disappointments may possibly be in store, I must yet insinuate to you that in a cause like this it would be criminal to feel discouragement. In the very design and endeavor to establish the Atlantic Telegraph there is almost enough of glory. It is true if it only be an attempt there would not be quite enough of profit. I hope that will come, too; but there is enough of public spirit, of love for science, for our country, for the human race, almost to suffice in themselves. However, upon the rocky frontlet of Ireland, at all events, to-day we will presume upon success. We are about, either by this sundown or by to-morrow's dawn, to establish a new material link between the Old World and the New. Moral links there have been—links of race, links of commerce, links of friendship, links of literature, links of glory; but this, our new link, instead of superseding and supplanting the old ones, is to give a life and intensity they never had before. Highly as I value the reputations of those who have conceived, and those who have contributed to carry out this bright design—and I wish that so many of them had not been unavoidably prevented from being amongst us at this moment—highly as I estimate their reputation, yet I do not compliment them with the idea that they are to efface or dim the glory of that Columbus, who, when the large vessels in the harbor of Cork yesterday weighed their anchors, did so on that very day 365 years ago—it would have been called in Hebrew writ a year of years—and set sail upon his glorious enterprise of discovery. They, I say, will not dim or efface his glory, but they are now giving the last finish and consummation to his work. Hitherto the inhabitants of the two worlds have associated perhaps in the chilling atmosphere of distance with each other—a sort of bowing distance; but now we can be hand to hand, grasp to grasp, pulse to pulse. The link which is now to connect us, like the insect in the immortal couplet of our poet:

> ——While exquisitely fine,
> Feels at each thread and moves along the line.

And we may feel, gentlemen of Ireland, of England, and of America, who may happen to be present, that we may take our stand here upon the extreme rocky edge of our beloved Ireland; we may, as it were, leave in our rear behind us the wars, the strifes, and the bloodshed of the elder Europe, and I fear I may say, of the elder Asia; and we may pledge ourselves, weak as our agency may be, imperfect as our powers may be, inadequate in strict diplomatic form as our credentials may be, yet, in the face of the unparalleled circumstances of the place and the hour, in the immediate neighborhood of the mighty vessels whose appearance may be beautiful upon the waters, even as are the feet upon mountains of those who preach the Gospel of peace—as a homage due to that serene science which often affords higher and holier lessons of harmony and good-will than the wayward passions of man are always apt to learn—in the face and in the strength of such circumstances, let us pledge ourselves to eternal peace between the Old World and the New. Why, gentlemen, what excuse would there be for misunderstanding? What justification could there be for war, when the disarming message, when the full explanation, when the genial and healing counsel may be wafted even across the mighty Atlantic, quicker than the sunbeam's path and the lightning's flash? I feel, gentlemen, that I shall best embody the sentiments which I am sure pervade this entire meeting—the sentiments most akin to this company and this hour, if, after having drunk the health of the gentle mistress of the British Islands, I now call upon you to drink, with congenial honors, to the lasting friendship of the British Islands and of America, and to the health and welfare of the President of the United States."

On the afternoon of Wednesday, August 5, the shore-end of the Cable was safely landed at Valentia. The Lord Lieutenant formally received it from Lieut. PENNOCK of the U.S. steamer *Susquehanna*, to whom the duty of the landing had been assigned. As his Excellency received it, he gave expression to a hope that the work so well begun would be carried to a satisfactory completion.

The scene in the harbor of Valentia at this time was extremely animated and exciting. The shore was covered with an immense multitude, attracted by the extraordinary interest of the occasion. The bay was dotted with vessels of all descriptions, filled with eager spectators of the scene. His Excellency the Lord Lieutenant was among the first to seize the end of the Cable, as it was passed on shore, and in a few moments the attachment was firmly made on the Irish coast, in the telegraph house at the head of Valentia bay.

The wire having been safely secured, the Reverend Mr. DAY, of Kenmore, pronounced the following prayer:

O Eternal Lord God, who alone spreadest out the heavens, and rulest the raging of the sea; who hast compassed the waters with bounds, till day and night come to an end; and whom the winds and the sea obey; Look down in mercy, we beseech thee, upon us thy servants, who now approach the throne of grace; and let our prayer ascend before thee with acceptance. Thou hast commanded and encouraged us, in all our ways, to acknowledge thee, and to commit our works to thee (Prov. iii. 5, 6; xvi. 3); and thou hast graciously promised to direct our paths, and to prosper our handi-work. We desire now to look up to thee; and believing that without thy help and blessing, nothing can prosper or succeed, we humbly commit this work, and all who are engaged in it, to thy care and guidance. Let it please thee to grant to us thy servants wisdom and power, to complete what we have been led by thy Providence to undertake; that being begun and carried on in the spirit of prayer, and in dependence upon thee, it may tend to thy glory: and to the good of all nations, by promoting the increase of unity, peace, and concord.

Overrule, we pray thee, every obstacle, and remove every difficulty which would prevent us from succeeding in this important undertaking. Control the winds and the sea by thy Almighty power, and grant us such favorable weather that we may be enabled to lay the Cable safely and effectually. And may thy hand of power and mercy be so acknowledged by all, that the language of every heart may be, "Not unto us, O Lord, not unto us, but unto thy name give glory," that so thy name may be hallowed and magnified in us and by us.

Finally, we beseech thee to implant within us a spirit of humanity and childlike dependence upon thee; and teach us to feel as well as to say, "If the Lord will we shall do this or that."

Hear us, O Lord, and answer us in these our petitions, according to thy precious promise for Jesus Christ's sake. Amen.

The Lord Lieutenant then addressed the assemblage, as follows:

"My American, English, and Irish friends, I feel at such a moment as this that no language of mine can be becoming except that of prayer and praise. However, it is allowable to any human lips, though they have not been specially qualified for the office, to raise the ascription of 'Glory to God in the highest; on earth peace, good will to men.' That, I believe, is the spirit in which this great work has been undertaken; and it is this reflection that encourages me to feel confident hopes in its final success. I believe that the great work now so happily begun will accomplish many great and noble purposes of trade, of national policy and of empire. But there is only one view in which I will present it to those whom I have the pleasure to address. You are aware—you must know, some of you, from your own experience—that many of your dear friends and near relatives have left their native land to receive hospitable shelter in America. Well, then, I do not expect that all of you can understand the wondrous mechanism by which this great undertaking is to be carried on. But this, I think, you all of you understand. If you wished to communicate some piece of intelligence straightway to your relatives across the wide world of waters—if you wished to tell those whom you know it would interest in their heart of hearts, of a birth, or a marriage, or alas, a death, among you, the little cord, which we have now hauled up to shore, will impart that tidings quicker than the flash of the lightning. Let us indeed hope, let us pray that the hopes of those who have set on foot this great design, may be rewarded by its entire success; and let us hope, further, that this Atlantic Cable will also, in all future time, serve as an emblem of that strong cord of love which I trust will always unite the British islands to the great continent of America. And you will join me in my fervent wish that the Giver of all Good, who has enabled some of his servants to discern so much of the working of the mighty laws by which he fills the universe, will further so bless this wonderful work, as to make it even more to serve the high purpose of the good of man and tend to His great glory. And now, all my friends, as there can be no project or undertaking which ought not to receive the approbation and applause of the people, will you join with me in giving three hearty cheers for it. Three cheers are not enough for me—they are what we give on common occasions—and as it is for the success of the Atlantic Telegraph Cable, I must have at least one dozen cheers."

Mr. Cyrus W. Field was called upon. He said:

"I have no words to express the feelings which fill my heart to-night—it beats with love and affection for every man, woman, and child who hears me. [Cheers.] I may say, however, that, if ever at the other side of the waters now before us, any one of you shall present yourselves at my door and say that you took hand or part, even by an approving smile, in our work here to-day, you shall have a true American welcome. I cannot bind myself to more, and shall merely say, 'What God hath joined together, let no man put asunder.'"

On the evening of Friday, the 7th of August, 1857, the Telegraphic Squadron bore away from the coast of Ireland, delivering the Cable into the sea at a slow and steady rate. The Company having decided upon the attempt to lay the Cable by commencing at the Irish shore, and effecting a splice in mid-ocean, the work of paying-out was begun by the *Niagara* alone. Unfortunately the commencement of the Expedition was inauspicious. When about four miles of the thick shore-end of the Cable had been payed out, it became entangled with the machinery, owing to a momentary want of watchfulness, and parted. An attempt was immediately made to recover the lost portion. The *Niagara* came to anchor for the night. On the following day, the Cable was recovered; a splice was made, and the work was resumed without further accident to the shore-end.

At noon on Sunday, August 9, ninety-five miles of Cable had been expended, the continuity of the electric current remaining perfect, and signals passing between the *Niagara* and the station at Valentia.

On Monday, August 10th, at 8 45 P.M., and for two

hours afterwards, the electricians failed to receive signals, the continuity being now, for the first time, interrupted. Towards midnight the current was re-established, but the hopes which this circumstance revived were of short duration.

On Tuesday, August 11th, at 3 45 A.M., the machinery stopped, and with the strain the Cable parted. Three hundred and forty-four miles of the Cable were lost; the depth of water in which it was submerged being about two miles.

The first Expedition having thus come to an untimely end, nothing remained but to return to Ireland. The *Niagara* accordingly put about, and headed for Valentia. The following letter from Engineer BRIGHT was publicly read on board the *Niagara* by Captain HUDSON, on the return trip. It expressly exculpates the officers and men of the *Niagara* from any responsibility for the disaster:

AT SEA, ON BOARD THE NIAGARA,
Thursday, Aug. 13, 1857.

SIR:—I feel it my duty before leaving the *Niagara* to state that I do not attribute the fracture of the Cable to be in the least degree attached to any one connected with the ship; on the contrary, I must take this opportunity of expressing, on the part of the Company, the great obligation which we are under to yourself, your officers and men, and I shall esteem it a favor if you will thank them, on our behalf, for the never-failing zeal and attention which has been so universally displayed in our cause.

I am, Sir, &c., &c., &c.

(Signed,) CHARLES T. BRIGHT,

Engineer to the Atlantic Telegraph Company.

To Captain HUDSON, U.S.N., &c., &c., &c.

It appears that at the time the Cable parted, there was a heavy swell in the sea, and that while the vessel was making some three or four knots an hour, the Cable was running out at the rate of five or six, and sometimes even seven knots. Mr. BRIGHT, believing that the quantity of wire provided for the Expedition would, at this rate, be exhausted before the Newfoundland shore could be reached, adopted a measure which unhappily proved fatal to the enterprise. On the afternoon previous to the accident, the Cable was thrown out of the controller on two different occasions, and suffered severe strains. These mishaps had given rise to gloomy apprehensions. The pressure upon the Cable was gradually increased until, at the time of the rupture, a force of 3,000 pounds was applied. At this moment the stern of the *Niagara* was low down in the trough of the sea. As the ship rose on the back of the waves, the extra strain thus occasioned was more than the strength of the Cable could bear. It gave way under the pressure, and, parting at some distance from the ship's stern, sank like lead.

On the afternoon of Friday, August 14th, the *Niagara* and the *Agamemnon*, having joined company, arrived at Plymouth, England, attended by the *Susquehanna*. Here the fleet awaited further orders. So much of the Cable had been lost, disasters had appeared to multiply with such marvellous celerity, the season was so far advanced, that a new attempt seemed nearly impracticable during that year, and the further progress of the Expedition became a matter of serious consideration. Grave errors in the management of the enterprise had been developed during this Expedition; the operation of the machinery for paying out the Cable was discovered to be defective; a suspicion was excited that the plan needed a radical alteration; the length of Cable provided by the Company had proved inadequate to meet unforeseen contingencies. In view of these circumstances, the necessity of a reconstruction and thorough modification of the programme became evident. The Directors, after a series of meetings, held at their office in London, finally resolved to postpone, but not to abandon the enterprise.

The following is Mr. BRIGHT'S official report of this disaster:—

LONDON, *Tuesday*, *Aug.* 18.

GENTLEMEN—I forwarded by the *Leopard* a brief statement of the circumstances attending the fracture of the Cable on the 11th inst., and I have now to lay before you the full particulars connected with the expedition.

After leaving Valentia on the evening of the 7th inst., the paying out of the Cable from the *Niagara* commenced most satisfactorily until immediately before the mishap.

At the junction between the shore and the smaller Cable, about 8 miles from the starting point, it was necessary to stop to renew the splice; this was successfully effected, and the end of the heavier Cable lowered by a hawser until it reached the bottom, buoys being attached at a short distance apart to mark the place of union.

By noon of the 8th we had paid out 40 miles of Cable, including the heavy shore end, our exact position at that time being in lat. 51° 59′ 36″ N., long. 11° 19′ 15″ W., and the depth of water, according to the soundings taken by the *Cyclops*, whose course we nearly followed, 90 fathoms.

Up to 4 P.M. on that day, the egress of the Cable had been sufficiently retarded by the power necessary to keep the machinery in motion at a rate a little faster than the speed of the ship; but, as the water deepened, it was necessary to place some further restraint upon it by applying pressure to the friction drums in connection with the paying-out sheaves, and this was gradually and cautiously increased from time to time as the speed of the Cable, compared with that of the vessel, and the depths of the soundings showed to be requisite.

By midnight 85 miles had been safely laid; the depth of water being then a little more than 200 fathoms.

At 8 o'clock in the morning of the 9th, we had finished the deck coil in the after part of the ship, having paid out 120 miles. The change to the coil between decks forward was safely made.

By noon we had laid 136 miles of Cable. the *Niagara* having reached lat. 52° 11′ 40″ N., long. 13° 10′ 20″ W., and the depth of water having increased to 410 fathoms.

In the evening the speed of the vessel was raised to 5 knots per hour. I had previously kept down the rate at from 3 to 4 knots for the small Cable, and 2 for the heavy end next the shore, wishing to get the men and machinery well at work prior to attaining the speed which I had anticipated making.

By midnight 189 miles had been laid.

At 4 o'clock in the morning of the 10th, the depth of water began to increase rapidly from 550 fathoms to 1,750 in a distance of 8 miles. Up to this time 7 cwt. strain sufficed to keep the rate of the Cable near enough to that of the ship; but as the water deepened the proportionate speed of the Cable advanced, and it was necessary to augment the pressure by degrees until in the depth of 1,700 fathoms, the indicator showed a strain of 15 cwt., while the Cable and ship were running 5½ and 5 knots respectively.

At noon on the 10th we had paid out 255 miles of Cable, the vessel having made 214 miles from shore, being then in lat. 52° 27′ 50″ N., long. 16° 0′ 15″ W. At this time we experienced an increasing swell, followed later in the day by a strong breeze.

From this period, having reached 2,000 fathoms water, it was necessary to increase the strain to a ton, by which the rate of the Cable was maintained in due proportion to that of the ship.

At 6 in the evening some difficulty arose through the Cable getting out of the sheaves of the paying-out machine, owing to the tar and pitch hardening in the grooves, and a splice of large dimensions passing over them. This was rectified by fixing additional guards, and softening the tar with oil.

It was necessary to bring up the ship, holding the Cable by stoppers until it was again properly disposed around the pulleys. Some importance is due to this event, as showing that it is possible to lie to in deep water without continuing to pay out the Cable—a point upon which doubts have been frequently expressed.

Shortly after this the speed of the Cable gained considerably upon that of the ship, and up to 9 o'clock, while the rate of the latter was about 3 knots by the log, the Cable was running out from 5¼ to 5¾ knots per hour. The strain was then raised to 25 cwt.; but the wind and sea increasing, and a current at the same time carrying the Cable at an angle from the direct line of the ship's course, it was not found sufficient to check the Cable, which was at midnight making 2½ knots above the speed of the ship, and sometimes imperilling the safe uncoiling in the hold. The retarding force was therefore increased at 2 o'clock to an amount equivalent to 30 cwt., and then again, in consequence of the speed continuing to be more than it would have been prudent to admit, 35 cwt. By this the rate of the Cable was brought to a little short of 5 knots, at which it continued steadily until 4 45, when it parted, the length paid out at that time being 335 miles.

I had, up to this time, attended personally to the regulation of the breaks; but, finding that all was going on well, and that it being necessary that I should be temporarily away from the machine to ascertain the rate of the ship, and to see how the Cable was coming out of the hold, and also to visit the electrician, the machine was for the moment left in charge of a mechanic who had been engaged from the first in its construction and fitting, and was acquainted with its operation.

I was proceeding to the fore part of the ship when I heard the machine stop; I immediately called out to ease the break and reverse the engine of the ship, but when I reached the spot the Cable was broken.

On examining the machine, which was otherwise in perfect order, I found that the breaks had not been released, and to this, or to the handwheel of the break being turned the wrong way, may be attributed the stoppage, and the consequent fracture of the Cable. When the rate of the wheels grew slower as the ship dropped her stern in the swell, the break should have been eased; this had been done regularly before, whenever an unusually sudden descent of the ship temporarily withdrew the pressure from the Cable in the sea; but, owing to our entering the deep water the previous morning, and having all hands ready for any emergency that might occur there, the chief part of my staff had been compelled to give in at night through sheer exhaustion, and hence, being short-handed, I was obliged for the time to leave the machine without, as it proves, sufficient intelligence to control it.

I perceive that on the next occasion it will be needful, from the wearing and anxious nature of the work, to have three separate relays of staff, and to employ, for attention to the breaks, a higher degree of mechanical skill.

The origin of the accident was no doubt the amount of retarding strain put upon the Cable, but had the machine been properly manipulated at the time it could not possibly have taken place.

It has been suggested as a cause of the failure that the machinery is too massive and ponderous. My experience of its action teaches otherwise; for three days in shallow and deep water, as well as in rapid transition from one to the other, nothing could be more perfect than its working, and since it performed its duty so smoothly and efficiently in the smaller depths, where the weight of the Cable had less ability to overcome its friction and resistance, it can scarcely be said to be too heavy for deep water, where it was necessary for the increased weight of Cable to restrain its rapid motion by applying to it a considerable degree of additional friction. Its action was most complete, and all parts worked well together. I see how it can be improved by a modification in the form of sheave, by an addition to the arrangement for adjusting the breaks, and some other slight alterations; but with proper management, without any change whatever, I am confident that the whole length of the Cable might have been safely laid by it, and it must be remembered, as a test of the work which it has done, that, unfortunate as this termination to the expedition is, the longest length of Cable ever laid has been paid out by it, and that in the deepest water yet passed over.

After the accident had occurred, soundings were taken by Lieutenant DAYMAN, and the depth found to be 2,000 fathoms.

It will be remembered that some importance was attached to the Cables in the *Niagara* and *Agamemnon* being manufactured in opposite lays. I thought this a favorable opportunity to show that practically the difference was not of consequence in affecting the junction in mid-ocean. We therefore made a splice between the two vessels, and several miles were then paid out without difficulty.

I requested the commanders of the vessels to proceed to Plymouth, as the docks there afforded better facilities than any other port for landing the Cable, should it be necessary to do so.

The whole of the Cable on board has been carefully tested and inspected, and found to be in as perfect condition as when it left the works at Greenwich and Birkenhead.

One important point presses for your consideration at an early period; a large portion of the Cable already laid may be recovered at a comparative small expense. I append an estimate of the cost, and shall be glad to receive your authority to proceed with this work.

I do not perceive in our present condition any reason for discouragement, but I have, on the contrary, a greater confidence than ever in the undertaking. It has been proved beyond a doubt that no obstacle exists to prevent our ultimate success, and I see clearly how every difficulty which has presented itself in this voyage can be effectually dealt with in the next.

The Cable has been paid at the expected rate in the great depths; its electrical working through the entire length has been most satisfactorily accomplished, while the portion laid actually improved in efficiency by being submerged, from the low temperature of the water, and the close compression of the texture of the gutta percha.

The structure of the Cable has answered every expectation that I had formed of it, and if it were now necessary to construct another line I should not recommend any alteration from the present Cable, which in its working has confirmed my belief that it is expressly adapted to our requirements. Its weight in the water is so adjusted to the depth that the strain is within a manageable scope, while the effect of undercurrents upon its surface proves how dangerous it would be to attempt to lay a much lighter rope, which would, by the greater time occupied in sinking, expose an increased surface to their power.

I have the honor to remain, gentlemen,
Yours very faithfully,
CHARLES T. BRIGHT.

Captain HUDSON'S official report to the Navy Department was as follows:—

UNITED STATES STEAM-FRIGATE NIAGARA,
PLYMOUTH, England, *Friday*, *Aug.* 14, 1857.

SIR—I have the honor, as well as the mortification, to report the arrival of the *Niagara* at this port, after having run out 334 miles of Cable, some portions of it in a depth of over 2,050 fathoms, or more than 2¼ miles, when it was broken by too much pressure on the brake attached to the machinery for paying it out. I have every reason to believe, from what we have thus far experienced in wire laying, that under ordinary circumstances of weather, and with machinery adapted to the purpose—for such as we have on board requires altering and improving—the Cable may be laid in safety on the track marked out for it over the Atlantic ocean.

At the time the Cable parted—Aug. 11, 3 45 A.M.—the ship was going along 4 knots, and had been running at the rate of from 3 to 4 knots through the night, with some motion from a moderate head sea, and the Company's chief engineer and men attending their brakes to lessen the expenditure of Cable, until they finally carried it away, which made all hands of us through the day like a household or family which had lost their dearest friend, for officers and men had become deeply interested in the success of the enterprise.

Mr. FIELD left the ship soon after the accident occurred, in H. B. M. steam-brig *Cyclops*, for Valentia Bay, Ireland, requesting that the *Niagara*, *Susquehanna*, and *Agamemnon* should proceed to this place, after making certain experiments with the wire and machinery in deep water. The *Leopard* proceeded at once to Spithead.

Whether the Company intend to supply additional Cable, and try it again this season, or defer it until next summer, I am as yet unadvised. *If the latter*, the wire will have to be taken out of the ship and *retarred*, to save it from the effects of rust. I presume a few days will solve their present difficulties as to further action; and if their effort is not to be renewed at this late season of the year, I shall require further instructions to govern my future proceedings with this ship.

I herewith inclose a copy of communications received from the Telegraphic Company, while at Queenstown, or Cove of Cork, Ireland; also the certificate or letter of their chief engineer, Mr. BRIGHT, exonerating all the officers and men connected with the *Niagara* from any accountability or blame in relation to the parting or loss of the Telegraphic Cable.

I am respectfully,

Your obedient servant,

WILLIAM L. HUDSON.

The operations of the Electricians' department, during this trip, having been superintended by Prof. MORSE, thai gentleman gave publicity to certain reflections upon the conduct of Chief Engineer BRIGHT. In a journal of the voyage, published with the sanction of Prof. MORSE, the following language was employed:

> Our ship was going at the rate of four miles and two fathoms per hour, and the Cable running out at greater speed, perhaps at the rate of five miles the hour. Mr. BRIGHT spoke to the man in charge of the brakes, asking him what strain was on the Cable, to which the answer was returned, "About three thousand pounds." Mr. BRIGHT directed him to put one hundred pounds more of force upon the brakes, to check the speed of the cable. This was demurred to by the man for a moment, who expressed a fear that it would not be prudent. Mr. BRIGHT persevered in his orders.

An ample retraction of this aspersion was subsequently made by Prof. MORSE, on the receipt of an explanatory letter from Mr. BRIGHT. This letter was published in the American papers, towards the end of October, 1857, at the request of Prof. MORSE. In the course of his explanatory statement, Mr. BRIGHT observes:

> I am quite willing to take the reproach to myself which always

belongs to a want of success in any enterprise, but will you allow me to correct your narrative? I had been on deck all night; the brakes had been regulated by myself or Mr. Clifford, one of the Assistant Engineers on board, the whole time. The strain which was on the Cable when it parted had been on for some time; I gave the man at the brake no orders to alter the adjustment, nor did he demur to any, nor make any such observation as you allude to. I set the brakes three-quarters of an hour, at least, before the accident, and watched the effect carefully, until I was obliged to leave the machine, for the first time in two hours, to visit the hold and the electrical room, and to ascertain the rate of the ship, as reported to the officers of the deck. Before I had been absent two minutes, the accident occurred.

My only reason in troubling you with this is to correct your impression that I persisted in increasing the strain when the men under my command hesitated.

I am, my dear sir,

Very truly yours,

Charles T. Bright.

Prof. S. F. B. Morse, *Poughkeepsie.*

It is right that the opportunity of exculpation should be accorded to Mr. Bright, in connexion with this history.

CHAPTER VI.

THE EXPEDITION OF 1858.

THE Directors of the Company, undismayed by the failure which had attended the initial attempt in the summer of 1857, immediately proceeded to revise their plan of operations, with the view of introducing such improvements as should render the ultimate success of their enterprise more certain. It was a natural effect of failure that the mass of the public, both in England and America, lost confidence in the practicability of the wonderful project, and that the Company should be called to experience the fate that invariably attends the movers in a new field. The Submarine Cable began to be looked upon by the disbelieving as a suspicious speculation, in which unlimited capital might be sunk, and innumerable hopes crushed, without the smallest chance for a profitable return. The magnitude of the work was regarded with pride; its projectors were accorded the meed of praise which cannot be withheld from the most unsuccessful, when it is once fairly understood that an honest faith has underlaid the attempt to bring forth a new work; and the utility of a means of instantaneous communication between two nations so closely linked as England and America was universally acknowledged. Yet the expression of distrust became general. The disasters which had attended the first attempt to lay the Cable, the enormous expenditure that had been incurred, the apparently insurmountable difficulties which remained to be encountered, all became powerful arguments in opposition to the new programme of the Company. Capitalists, usually eager to embark in gigantic enterprises, were unwilling to venture investments in an operation that promised distant returns, or none at all. The tone of the Press, with a few sagacious exceptions, was lukewarm, if not absolutely hostile. The progress of the work was regarded with disfavor by all parties, with the exception of the small number of determined gentlemen who constituted the Direction of the Company, whose confidence remained firm, and whose energy knew no rest. To these gentlemen, among whose names that of Mr. CYRUS W. FIELD occupies an honored place, the final triumph seemed a matter of easy accomplishment,—easy, because wisdom had been gained by experience,—easy, from the fact, that the dangers already developed in the course of careful experiment carried with them the suggestion of methods for overcoming them.

Accordingly, the Directors began their preparations early in the fall of 1857. A general invitation was extended to scientific gentlemen to furnish the results of their experience, as guides to future operations. The assistance of skilful mechanics was invited. The coöperation of adepts in matters of scientific importance, and in the details of mechanical arrangement, was secured, and the plan of 1858 began to assume definite forms before the end of 1857. Frequent conferences, held at short intervals from October to December, 1857, resulted in the adoption of a number of important modifications, which may be generally indicated under four divisions, viz.:

1. A junction of the Telegraphic Cable in mid-ocean.
2. The provision of a greater length of Cable.
3. The selection of an earlier season of the year.
4. An improvement in the paying-out machinery.

The reasons for these changes appeared irrefutable. It was argued that the Submarine Cable would be less liable to breakage if paid out from mid-ocean, than if it were started from one end of the line and taken up by the second vessel at a point equidistant from the two shores. A greater length of Cable was ordered, to provide against the contingency of fresh mishaps. The month of June was regarded as a more favorable season of the year than August. The paying-out machinery having proved clumsy and inefficient upon the first trial, underwent essential changes, and was rendered as nearly perfect as mechanical ingenuity could make it. These were the radical alterations which entered into the management of the second Expedition.

While the Company were engaged in perfecting the arrangements for a new Expedition, efforts were made, under the direction of their Chief Engineer, Mr. BRIGHT, to recover the submerged portion of the Cable which was lost at the time of the first failure. The British steamer *Leipsic* was detailed for this service. Operations were commenced Oct. 22, 1857. The Cable had been secured to the Irish shore by a heavy shore-line, made fast in the station at Valencia. An ingenious apparatus was constructed for under-running this shore-cable to the point of junction with the main line. A heavy frame of timber (technically, a *catamaran*), bearing a saddle fixed between two iron buoys, was run under the shore-end, and then towed out by the *Leipsic*. The operation of hauling in and recoiling was successfully prosecuted until some fifty-three miles were recovered, when, the weather becoming boisterous, with heavy gales, the work suddenly ceased, from the parting of the wire, and this attempt ended. The part of the Cable which was recovered during this expedition bore no appearance of injury, and was found available for use. None of the gutta-percha coatings were disturbed, and the tarring of the wire remained perfect.

While these movements were in progress, the frigates *Niagara* and *Agamemnon* discharged the Cable at Keyham Docks in England. The portion shipped on board the latter vessel was discharged very slowly, at a rate of speed not exceeding one mile per hour. The share consigned to the *Niagara* was unshipped with great rapidity, in order to permit the return of that vessel to the United States. As the Cable was paid out, it passed through a composition of tar, pitch, linseed oil and beeswax, and was coiled, in compact circles, in tanks constructed for the purpose. Here it remained undisturbed until again shipped on board the *Niagara* and *Agamemnon* for the last Expedition. The *Niagara* returned to this port in the winter, and underwent examination, but was found to be in excellent condition, requiring but slight repairs.

In the course of the winter, the Company applied to the British and American Governments for the re-employment of the vessels of the squadron of 1857, in the new attempt resolved upon for the summer of 1858. To the credit of both Governments, this request was met by a cordially affirmative response. The British Government again placed the fine ship *Agamemnon*, Captain PRIDDIE, with the steam-tenders *Valorous* and *Gorgon*, under the direction of the Company. Our own Government, with commendable promptitude, re-assigned the *Niagara* to a similar service. The Secretary of the Navy wrote to the Directors, in December, 1857, as follows:—

NAVY DEPARTMENT, *Dec.* 30, 1857.

GENTLEMEN—Your communication of the 23d inst. has been received. I have to inform you, in reply, that by direction of the President of the United States, the steam-frigate *Niagara* will again be detailed for the purpose of assisting in laying the Telegraphic Cable next Summer. The Department will, agreeably to your request, give Chief-Engineer EVERETT leave of absence, with permission to leave the United States, that the Telegraphic Company may avail itself of his services in connexion with this work.

I am, respectfully, your obedient servant,

ISAAC TOUCEY.

The *Niagara* was again placed under command of Capt. W. L. HUDSON, U.S.N.—a most capable and energetic officer, to whose unwearied exertions and unbounded enthusiasm in the work, no small share in the final success of this enterprise is due. The *Niagara* sailed from the port of New York on the 9th of March, 1858, on her return to England, and arrived at Plymouth on the 24th of the same month.

With the opening of the Spring, the Company began active preparations for the new Expedition. Pending the negotiations with the Governments for the use of the Telegraphic Fleet, additional supplies of the Cable had been manufactured in England, the new paying-out machines were already under way, and all things were ordered to be in readiness for a second attempt early in the ensuing Summer. A description of the new machinery is embodied in another page of this work.

The re-shipment of the Cable for the Second Expedition was commenced at Keyham Docks (Plymouth, Eng.) on Friday, March 19, 1858. At first, the *Agamemnon* alone received the wire; the *Niagara* not having been fully prepared for the stowage of her portion. Various important alterations were made on board both vessels, with a view, not only of insuring greater safety in paying-out, but in order to accommodate the enlarged bulk of the Cable. The preliminary trial had demonstrated the existence of serious defects in the manner of shipment on board the *Agamemnon*. These errors were corrected. Instead of coiling the Cable in an oval form in the hold of that vessel, it was arranged in a circle, winding around a huge circular cone, twelve feet three inches high, ten feet in diameter at the base, and five feet at the apex. It was the breaking of this cone which afterwards so seriously imperilled the *Agamemnon*, during the heavy storms she encountered in June. A new coil was laid upon the upper deck, abaft the foremast; and another on the orlop deck. A new guard was also fixed at the stern of the ship, protecting the propeller, in order to prevent contact with the Cable during the process of paying-out.

The labor of coiling into the *Niagara* was accomplished at the rate of 30 miles per day; a portion was stored in the ward-room circle, and a portion in the lower forward hold, the work going on day and night without intermission. The ward-room circle had 311½ miles of the Cable coiled round a cone in the centre of the circle, the diameter of which was 38 feet, the cone itself being 9 feet diameter at bottom, by 4 feet 6 inches at the top, and extending to within 2 feet 6 inches of the deck. Around the cone were suspended three iron-bound bar hoops, about 18 inches from the cone, intended to guide and direct the Cable out as it passed from the tier. In this tier there were 89 flakes, laid down with great care, each flake averaging 270 turns round the cone, and there were employed constantly in the circle, to receive and coil away, 40 men, 30 of whom were of the crew, and 10 Company's men. As these men had to be relieved every four hours, eighty and sometimes more, were detailed for the duty in that circle, while at the same time another gang was similarly employed in the lower hold, the forward gang, consisting of 50 men from the ship, and 10 Company's men.

The lower hold differed very little, if any, from the ward-room circle or cone, except in being higher. In this part of the ship there were 359½ miles of cable. Immediately above this lower forward cone was the orlop-deck circle, the span and diameter the same as in the ward-rooms.

The third circle was immediately over the orlop-deck circle, with the same dimensions as the others. The berth deck received its portion of 200 miles. Directly over this circle was the upper-deck circle. Here 250 miles were coiled. Its diameter was 27 feet 6 inches.

There was still another circle on the quarter-deck circle. It was somewhat smaller, and contained 150 miles only. All the circles and cones were fitted the same in every respect, with great precision.

In addition, a massive structure 11 feet high, 20 feet long, and eight feet broad, was placed on the quarter-deck, intended to sustain machinery. Besides this, and independent of the coils of Cable in the circles (in the aggregate some 1,600 miles of Cable) a large portion of the old Cable lost last year and recovered was placed on the forecastle deck, being put on board for experimental purposes.

The presence of a large force of men was rendered necessary on board the ships, in order to effect a proper arrangement of the Cable and insure its safety during the process of paying-out. Thus the company on board the *Niagara*, who were actively engaged in the operations, consisted of a director, one or two superintendents, several cable-inspectors, four or five overseers, not less than six or eight electricians; with carpenters, black-smiths, cable-coilers, and 40 to 50 of the crew constantly employed;

and when two gangs were at work the complement was doubled. There was also, one of the ship's officers constantly on hand by day and by night, whose relief took place as regularly as the men's.

Among the new contrivances was an immense iron shield, called a "bird-cage," extending from quarter to quarter around the rudder, intended as a guard or protector from the cable fouling the rudder or propeller, as it was paid out from the ship.

The appearance of the *Agamemnon*, after receiving her portion of the Cable, was described as follows, by a correspondent of the *London Times*, who let no opportunity escape for disparaging the *Niagara:*

"Both the *Agamemnon* and *Niagara* are astonishingly deep. The lower deck ports of the former are very near the water, and they are being fastened and caulked before starting. But in spite of this, the *Agamemnon* carries her share infinitely better than her long black-looking rival of the United States, which is immersed very deeply indeed by her load. The *Agamemnon* only draws 26 feet, or actually one foot less than her draught at starting last year; but even at this depth she bears herself well, and looks a noble ship, and one that should be seaworthy in any weather. The *Niagara*, however, draws no less than 27 feet 2 inches aft, and this great draught effects a marvellous and most unpleasant change in her appearance, since it leaves her spar deck scarcely

8 feet above the water's edge. In fact, the main deck is actually below the water's level, and if her lofty bulwarks, some nine or ten feet high, were taken away, she would appear to be almost the last vessel in the world in which it was desirable to venture across the great Atlantic."

The stowage of the Cable on both ships, conducted slowly and with great care, occupied several weeks, and was completed in the early part of May. In addition to seven hundred miles of new Cable provided by the Company, condemned wire was shipped for the purpose of undertaking preliminary deep-sea experiments; so that the total length of Cable on board both ships on the 18th day of May was 3,008 miles, distributed as follows:

		Miles.
Niagara.—Good Cable	1,488	
Experimental Cable	22	—1,510
Agamemnon.—Good Cable	1,477	
Experimental Cable	17	—1,494
Aggregate.—Good Cable		2,965
Experimental Cable		39
Total		3,004

The new paying-out machines were placed on board the *Niagara* and *Agamemnon* while they lay at Plymouth, receiving the Cable. It is proper to state that the improved plan upon which these machines were constructed, belongs to Mr. HIRAM BERDAN, of New York, who furnished the Company with a model, a gentleman of remarkable scientific ability, widely known as the inventor of various important mechanical appliances. The manufacturers of the machines were Messrs. EASTON and AMOS, Southwark.

The principle of the operation of the Paying-Out Machine was simple. The whole of the important part of this apparatus consists of APPOLD'S self-regulating brake, which is so adjusted and constructed as always to exert a certain amount of resistance, which can be regulated by the revolution of the wheels to which it is applied. More than this fixed amount of resistance, whatever it may be, it cannot produce, no matter whether the machine is hot or dry, or covered with sand; and neither can it be worked at less than this amount, no matter to what extent all the friction surfaces of the wooden brake itself may be oiled. This well-known brake was first exhibited in the Great Exhibition of 1851, in the new labor machine constructed for prisons, in order to insure a certain amount of work from each convict. For this hard-labor purpose the brake is still extensively employed. It is made of bars of wood laid lengthwise across the edge of the wheel, over which it laps down firmly, and to which it is held with massive weights fixed to the ends of levers. It is the number and size of these weights which regulate precisely the degree of resistance to the revolutions of the wheel, and which, of course, enable those in charge of the machine to fix the pressure of the brake at what they please, and when so fixed nothing can alter it. In the new Telegraphic apparatus, this brake is attached over two drums connected with the two main grooved wheels, round which the actual Cable passes in running out. The latter are simply broad, solid, iron wheels, each cut with four very deep grooves, in which the Cable rests, to prevent it flying up or "overriding." It passes over these two main wheels, not in a double figure of eight, as in the old ponderous machine of four wheels, but simply wound over one, to and round the other, and so on four times, till it is finally paid down into the water. Thus, the wire was wound up from the hold of the vessel, passed four times over the double main wheels, connected with the brake or friction drums, past the register which indicated the rate of paying out and the strain upon the Cable, and then at once into the deep. The strain at which the Cable breaks is 62 cwt., and to guard against any chance of mishap, not more than half this strain was to be put upon it. The brakes, as a rule, were fixed to give a strain of about 16 cwt., and the force required to keep the machine going, or about 8 cwt. more, was the utmost that was to be allowed to come upon the wire.

Thus, therefore, the force required to sever the cable could never be exerted even by accident or mishap, no matter who might be in charge of the machine, nor how much the vessel might pitch and roll. The brake of the paying-out machine used on the occasion of the first attempt, was capable, by a movement of the hand, of exerting the most prodigious resistance to the turning of the wheels, and this formidable invention was used with so little care, that not until the injury was irreparable was the danger seen. The chief beauty, however, of the new machine was that, while nothing could add to the fixed strain of the brakes, any one could in a moment ease them as much as might be considered necessary, and until, in fact, there was no resistance at all beyond the 8-cwt. strain on the wire, which, as we have said, was required to keep the machine turning. So simple was the operation, that a child could remove the whole resistance of the brake and put it on again as often as 20 or 30 times a minute.

For this purpose, at a few feet from the paying-out machine, the Cable passed over a wheel which registered precisely the strain in pounds at which the coil was running out. Facing this register, was a steering wheel, precisely similar to that of an ordinary vessel, and connected in the same way with compound levers, which acted upon the brake. Thus the officer in charge of the apparatus stood by this wheel, and watched the register of strain or pitch of the vessel, opened the brakes by the slightest movement of his hand, letting the cable run freely as the stern rose. The same officer, however, could not by any possible method increase the actual strain on the cable, which remained always according to the friction at which the break was at first adjusted by the engineer.

The value and simplicity of the whole apparatus were made so manifest that it was evident, as far as the paying-out machine was concerned, all that mechanical skill could effect in aid of the great undertaking had been accomplished. The *Niagara* and the *Agamemnon* were each fitted with one of these machines, which, when operated by steam, could be used for the purpose of under-running or drawing back the Cable in case of any hitch rendering such operation necessary.

The *Niagara* and *Agamemnon*, having been fully prepared for the service, sailed from Plymouth for Queenstown, Ireland, on Saturday, May 29; and on the same day put to sea from the latter port, to undertake an experimental trip for the purpose of testing the Cable.

On the 31st of May, when in latitude 47° 12′ north, longitude 9° 32′ west, the depth of water being 2,530 fathoms, a series of deep-sea experiments was commenced. The *Niagara* and *Agamemnon* were connected by hawsers, stern to stern, distant from each other some twelve hundred feet. The Cable was paid out and spliced on board the *Agamemnon*, and the first experiment began. Two miles of Cable were paid out, when the wire parted. On the following day (Tuesday, June 1,) the Cable was re-spliced, and three miles were paid out; but in the attempt to haul in, the wire again parted. On Wednesday, June 2, the Cable was again spliced, but in a few minutes parted on board the *Agamemnon*. These experiments having been continued during three days and one night, ceased with this last attempt, and, after various trials of the operations of splicing, lowering and heaving-in the wire, the squadron set sail for Plymouth, whence reports of the results were forwarded to the Directors of the Company.

The following is Capt. HUDSON'S official report of the results of this trip:—

UNITED STATES STEAM FRIGATE NIAGARA,
PLYMOUTH SOUND (ENG.), *June* 3, 1858.

SIR,—I have the honor to report that the Telegraphic Squadron, consisting of the *Niagara*, and H.M. ships *Agamemnon*, *Valorous*, and *Gorgon*, put to sea from Plymouth Sound at 5 P.M. on the 29th ult., and proceeded to lat. 47° 12′ north, and long. 9° 32′ west, when we hove to, and the *Gorgon* obtained two casts of soundings with her deep-sea apparatus, and found the depth of water to be twenty-five hundred and thirty fathoms. We immediately commenced our experiments by hanging the *Niagara* and *Agamemnon* together by hawsers, stern to, and distant from each other some twelve hundred feet. The telegraphic wire on both ships was spliced together on the *Agamemnon*, and the Cable lowered down by the new machinery of Mr. Everett, until the bight of it was laid on the bottom of the ocean. Some additional wire was paid out, and in this position the electric current was found perfect through the whole length of the Cable, about three thousand miles.

In our various experiments of splicing, lowering down, and heaving in the wire, this ship and the *Agamemnon* were several times tied together stern on. We have also had a fair test of Mr. Everett's machinery for lowering and heaving up the Cable from the ship, as well as running it out under a speed of five or six knots. Indeed, our experiments, occupying three days and almost an entire night, have, I think, entirely settled some mooted points in relation to the electric current passing freely at great depths under the ocean.

We hope to have in our additional forty miles of new wire, to coal ship, and be ready to leave this port with the squadron, on the 10th inst., for our great work. The officers and crew enjoy good health.

I am, respectfully, yours, &c.

W. L. Hudson, Captain.

Hon. I. Toucy, Secretary of the Navy.

The report of Chief Engineer Everett was as follows:—

United States Steam Frigate Niagara,
At Sea, *Thursday, June* 3, 1858.

Cyrus W. Field, General Manager of the Atlantic Telegraph Company:

Sir,—For the information of yourself and the Directors, I submit the following statement of experiments made during this trip.

Monday, 4 p.m., May 31, lat. 47° 12′ N., long. 9° 32′ W., soundings 2,530 fathoms, this ship and the *Agamemnon* being attached stern to stern by a hawser, 180 fathoms of Cable were veered out for the end to be taken on board that ship to be spliced. At 5½ o'clock, signal being made "All ready," in accordance with previous arrangement, one mile of Cable was veered out. We then commenced hauling it in. At 6½ six o'clock had recovered half a mile, when Mr. Bright's message was received saying he desired to make a new splice. At 9 40 received message "All is ready," and again commenced paying out as before. At 10 34 P.M. two miles were out. After this amount was paid out, the strain upon the cable was 3,600 to 4,100 pounds. At 11 28 commenced hauling in, but very slowly, as the strain nearly approached the breaking point of the rope. At 11 45 the hawser securing the ships together parted on the *Agamemnon*, but the ships were retained nearly in the same relative positions by working the engine when required. At 1 40, having hauled in one mile, 506½ fathoms, the continuity was reported broken. We continued to haul in until 2 15, when the end came, having lost of the two miles paid out 110 fathoms.

On Tuesday, at 8 40 A.M., the ships having been secured and splice made as before, a quarter of a mile was paid out, hawser released and ships started ahead slowly, at the same time the Cable was allowed to run quite freely until two miles had been paid out, when a gradual restraint was applied until an additional one mile, 387 fathoms, had been paid out, making in all three miles, 387 fathoms. At this time (10 23) commenced hauling in, and had recovered 190 fathoms when the Cable parted. At 4 44 P.M., the two ends of the new Cable having been spliced, we paid out two and a half miles at a rate which had been previously agreed upon, the electricians passing signals through the whole length of Cable. At 6 15 P.M., the *Agamemnon* made signal the Cable was parted. We at once commenced hauling the strain running up to 5,100 pounds during the receiving of the first quarter of a mile. At 9 20 the end came in, having lost 80 fathoms on the two and a-half miles paid out.

Wednesday, June 2, at 7½ o'clock, experimental Cable was again spliced, one quarter of a mile paid out, hawser released, and the ships started ahead. In a few minutes the *Agamemnon* made signal Cable parted. We continued to pay out until three and a quarter miles were out. The ship was then backed—large buoy and watch buoy attached to the Cable. Ship again run ahead, and when three hundred fathoms had been paid out, the Cable parted on the machinery. The ship then made for the buoy, with the hope of recovering the end of the Cable; but while hauling in the watch buoy, the large buoy suddenly fell over, showing that it had separated from the Cable. Upon recovering it we found the rope stopper (3½-inch rope) had been cut off by the Cable. At 12 55, by the request of Mr. Woodhouse, we paid over the end of experimental Cable to ascertain how rapidly it could be run off the coil with safety, but no greater speed was attained than seven knots, as the Cable was being often stranded on the machine by the accumulation of tar in the grooves, which was so hard that no scraper could be made to remove it at any speed. All the Cable used to-day was that brought from Greenwich expressly for experimenting and was long since condemned. Undoubtedly it has been much exposed to the weather, and stowed where considerable sand or dirt has been thrown upon it. With the Cable which was recovered last year, and used by us trying the experiment, we had no serious difficulty in keeping the tar out of the grooves, it being comparatively soft, though the amount was beyond what I could have believed. The amount of tar upon this Cable is much greater than that upon the Cable intended to be laid down; therefore I believe we can make such provision as that it shall not become a serious obstacle.

The result of this experimental trip has demonstrated that we have the capability of hauling in the Cable to a greater extent than I had expected. Not that I believe any great distance could be recovered, but in the general depth of water where the Cable is to be laid, in good weather, should a fault go overboard before the ship could be stopped, I am of opinion sufficient of the Cable may be hauled in to remedy the fault.

The operation of the machinery generally is certainly satisfactory, and there is no alteration I can suggest other than in the tar-scrapers, which will require modifications. The amount of tar accumulating is so much beyond what could have been expected from last year's experience, owing to the repeated coatings it has received since it was unloaded from this vessel last October, that extraordinary provision will be required. As regards the attaching of buoys, we can attach them, but at a great risk of breaking the Cable, and they should not be used in deep water, except as a last resort.

The arrangements for coils, provisions for leading the rope, and all the other many particulars incidental to this work, which have been under the direction of Mr. Woodhouse, do not require any alteration, and fully meet the requirements. I am, respectfully, your obedient servant,

W. E. Everett.

But a single week elapsed from the return of the fleet from this trip, when the Expedition set sail from Plymouth for the second great ocean trial.

A distinguished Russian naval officer sailed in the *Niagara*, the reason of whose visit is explained by the following letter:—

LEGATION OF THE UNITED STATES,
ST. PETERSBURG, *Monday, March* 22, 1858.

CAPTAIN,—I have the honor of making you acquainted with the bearer of this letter, Lieutenant the Baron BOYE, Aid-de-camp to his Imperial Highness the Grand Duke CONSTANTINE.

Although he will need no special recommendation from this quarter, there is a duty devolving on me in the present instance which makes it proper that I should briefly notice the object of his visit to the *Niagara.* It is doubtless known to you already that the President of the United States, on the application of Mr. STOECKL, the Russian Minister at Washington, has consented that one officer of the Imperial Navy of Russia, such a one as his Imperial Highness the Grand Duke CONSTANTINE might be pleased to designate, should go on board the *Niagara* to witness the laying of the electric Cable between Europe and America.

In consequence of that consent, which I have reason to believe was most cheerfully given, Lieutenant BOYE has been designated by his Imperial Highness the Grand Duke CONSTANTINE, for the purpose which you have before you. It has, therefore, become my duty, in accordance with the advices from the Department of State, that I should furnish the officer with a suitable letter of recommendation to the Commander of the *Niagara.*

The duty thus incurred may be considered, perhaps, as already fulfilled in the statement I have just made, nothing more being required, I am sure, by way of insuring Lieutenant BOYE the reception to which he is justly entitled.

The circumstances which have led to the selection of Lieutenant BOYE for the mission with which he is charged, should justify the addition of a few remarks before closing my letter. You will not fail to see in the arrangement I have herein briefly noticed, a gratifying proof of the friendly relations which subsists between the Government of his Imperial Majesty, the Emperor of Russia, and that of the United States. Such a view would, of itself, enhance the pleasure of receiving on the deck of your vessel an officer of the Imperial Navy.

Russia, though at some distance from the shores of the great ocean which it is to be hoped may soon be connected by the electric Cable, is not indifferent to the sublime work in which you are about to engage. She sends as her representative, an officer of rank and abilities, one high in the confidence of his distinguished chief, the Grand Admiral of the Russian Navý, that he may be witness to an undertaking which, if successful, will be hailed with joyful satisfaction throughout the civilised world.

In recommending Lieutenant the Baron Boye, to your kind consideration, I not only carry into effect the wishes of President Buchanan on the subject, but at the same time discharge a duty in the highest degree agreeable to myself.

Your position in the naval service of our country must have given you opportunity for learning something of the courtesies which officers of the United States Navy have been accustomed to receive from those of the Russian Navy, when the ships of the two nations have come together on distant seas; and no one, I may say, Captain, more fully appreciates the value of such courtesies, or knows better how to return them than yourself.

Respectfully,

Your obedient serv't,

Thos. H. Seymour.

To Capt. W. L. Hudson, U.S. Steamer *Niagara.*

On Thursday, June 10, the entire Telegraphic Fleet steamed out of Plymouth Harbor. The Squadron consisted of four vessels—the United States steam frigate *Niagara*, with H.B.M. paddle-wheel steamer *Valorous* as tender; H.B.M. steam-frigate *Agamemnon*, with H.B.M. paddle-wheel steamer *Gorgon* as tender.

The Government of England detached two naval engineers from actual duty, for the purpose of assisting in taking charge of the machinery on board the *Agamemnon*. Mr. AMOS, of the firm of EASTON & AMOS (who manufactured the machinery), and who had given a great deal of time and attention to the interests of the undertaking, was also present on board the *Agamemnon*, for the purpose of assisting in the regulation of the machinery.

The arrangements on board the *Niagara*, were as follows: Messrs. EVERETT & WOODHOUSE were in charge of the operations, with Captain KELL as an assistant, and Messrs. FOLLANSBEE and MCELWELL in charge of the machinery.

After having been three days at sea, the Expedition was overtaken by a fearful gale, which continued without intermission for nine days. On the seventh day of this heavy weather, the ships, which continued to keep together, had to part company, and the *Agamemnon* was obliged to scud before the wind for thirty-six hours; her coals got adrift, and a coil of her Cable shifted, so that her captain for some time entertained serious apprehensions for her safety, and from the immense strain her waterways were forced open, and one of her ports was broken. Two of the sailors were severely injured, and one of the marines lost his reason from fright. Yet such was the consummate skill, good seamanship, and intrepidity of her commander, Captain PRIDDIE, that he was enabled to bring her to the appointed rendezvous, lat 52° 2′, long. 33° 18′. The *Niagara* rode out the storm gallantly, having only carried away her jib-boom and one wing of the figure-head, the American Eagle. The results of this severe gale on board the two frigates showed the gross injustice that had been done the *Niagara* by the English writer, whose remarks about her before she left Plymouth, we have quoted.

All the vessels having at length arrived at their central point of junction, the first splice of the Cable was made on the 26th. After having paid out two and a half miles each, owing to an accident on board the *Niagara*, the Cable parted. The ships having again met, the splice was made good, and they commenced to pay out the Cable a second time; but after they had each paid out forty miles, it was reported that the current was broken, and no communication could be made between the ships. Unfortunately, in this instance, the breakage must have occurred at the bottom, as the electricians, from the fine calculations which their sensitive instruments allowed them to make, were able to declare such to have been the fact, even before the vessels came together again. Having cast off this loss, they met for the third time, and recovered the connexion of the Cable on the 28th. They then started afresh, and the *Niagara*, having paid out over one hundred and fifty miles of Cable, all on board entertained the most sanguine anticipations of success, when the fatal announcement was made upon Tuesday, the 29th, at 9 P.M., that the electric current had ceased to flow. As the necessity of abandoning the project for the present was now only too manifest, it was considered that the opportunity might as well be availed of, to test the strength of the Cable. Accordingly, the *Niagara*, with all her stores, was allowed to swing to the Cable, and, in addition, a strain of four tons was placed upon the brakes, yet, although it was blowing fresh at the time, the Cable held her as if she had been at anchor, for over an hour, when a heavy pitch of the sea snapped the Cable, and the *Niagara* bore away for Ireland. Before starting, an arrangement was made that should any accident occur in giving out the Cable before the ships had gone one hundred miles, they were to return to their starting-place in mid-ocean; but that, in case that distance should have been exceeded, before any casualty happened, they should make for Queenstown. In accordance with this understanding, the *Niagara*, having made one hundred and nine miles before this mishap, returned to Queenstown, arriving July 5. The events of this unsuccessful cruise are recorded, in connected form, in the following

JOURNAL OF THE VOYAGE OF THE NIAGARA.

Thursday, June 10.—At 11 A.M. cast off from the mooring-buoy in Plymouth Sound, and proceeded to sea in company with the *Agamemnon*, *Valorous*, and *Gorgon*. Weather fair; at night, nearly calm.

Friday, June 11.—Calm, and a smooth sea. Captain HUDSON sent a boat to the *Agamemnon*, tendering a tow, which was declined. The Captain of the *Agamemnon* stated, that he would accept the offer if the light breezes held a day or two longer. The progress made this day was very slow, the *Niagara* continually stopping for the Squadron to come up. Weather calm; lat. 49° 12′; long. 6° 53′.

Saturday, June 12.—Light N.N.E. winds and pleasant; the *Agamemnon*, *Valorous*, and *Gorgon* made all sail possible; the *Niagara* set no sails. Lat. 49° 42′; long. 10° 12′.

Sunday, June 13.—Commenced clear, but soon became squally; wind hauled to S.S.W. The *Niagara* set topsails and shut off steam. The wind freshened and rain set in. Lost sight of the *Valorous* and *Gorgon*. The wind increasing, the *Niagara* triple-reefed topsails and reefed foresail. The *Agamemnon* was at this time one mile distant. Lat. 50° 11′; long. 13° 17′.

Monday, June 14.—Squally, rainy weather; wind S.W., barometer fell from 30.35 to 29.17; wind increased to a gale, with a high sea. The *Niagara* close-reefed topsails, furled mizen-topsail, set storm fore-staysail, and furled foresail, keeping within one and a half miles of the *Agamemnon* throughout. Lat. 50° ·22′; long. 15° 57′.

Tuesday, June 15.—Strong gales from S.S.W., the *Niagara* under easy sail; at 4 A.M. wind moderated, and ship made more sail. Before night the wind again increased, and the ship was put under close-reefed fore and maintopsails, with storm fore-staysail. Lat. 51° 22′; long. 18° 47′.

Wednesday, June 16.—Strong gales; the *Agamemnon* in company; the *Valorous* and *Gorgon* out of sight. In the latter part of the day wind died away, and the *Niagara* made more sail. Distance sailed, by log, 135 miles, on N.W. by N. course. Day ended misty and foggy.

Thursday, June 17.—Weather still foggy, with stiff breeze; the *Agamemnon* in company; exchanged signals. During the day the *Niagara* passed a ship's boat, bottom up. The speed of the ship was five knots an hour, with small head of steam and little sail. Discovered a strange sail to the southward. Lat. 52° 35′; long. 23° 16′. One week at sea.

Friday, June 18.—The day commenced clear, with strong breezes from S.W. by S.; the ship under easy sail and but little steam; the *Agamemnon* in company;

the *Valorous* and *Gorgon* invisible; lat. 53° 18′; long. 25° 49′.

Saturday, June 19.—Weather overcast; wind strong; signaled a clipper-ship bound west. This day all hands on board the *Niagara* were mustered, and the Articles of War were read to the officers and crew. At noon strong winds from W.S.W. Lat. 54° 23′; long. 27° 50′.

Sunday, June 20.—Heavy gales and a high sea, the ship under very short sail; the *Agamemnon* one and a half miles distant, laboring terribly; from 4 A.M. to noon gale increases in violence; the squalls come more heavy and more frequent; the *Agamemnon* telegraphs, "We are going to wear ship;" she wears round on other tack; wind W.S.W.; the *Niagara* does the same; the *Agamemnon* again telegraphs, "We have lost our stern-guard;" the *Niagara*, at this time rolling heavy, brings the large iron buoys lashed outside under water; the lashings to the starboard buoy part, and carry away the cranes which support the buoys on each quarter. It is found necessary to cut adrift the starboard one and let it drift away; the port buoy is with difficulty secured and taken on board; at 4 P.M. the A. signalises, "We shall wear ship;" both ships rolling heavy, A. more especially; two strange vessels in sight; ends heavy gales; the barometer falls down to 29.19; at noon lat. 54° 12′, and long. 29° 36′ W.

Monday, June 21.—Heavy gales, high sea,. but clear

weather; wind S.W. by S.; at 6½ A.M. a very heavy sea strikes the jib-boom, which carries away the flying jib-boom and all attached to it. The sail and riggers, however, are saved. The *Agamemnon* distant one and a half miles; she is observed to have a heavy list, laboring very heavy; she is suddenly lost to our view; we conjecture she must have wore ship without signalising. At noon there is no prospect of its abating; the *Niagara* hove to, no observation; lat. by account 54°; 25° N. long., barometer 29° 21′, air 53°, water 51°. From noon to 6 P.M. blows the same; at 8 P.M. it moderates, and at midnight but little wind, and the sea becomes smooth.

Tuesday, June 22.—Fine weather, smooth sea, wind moderate from the westward; chased a ship bound west, mistook her for the *Agamemnon*. Noon furled all sail, the wind light and dead ahead; lat. 53°, 42′, long. 30° 17′ west. From noon to midnight same variation of compass 3½ points W.; local attraction 23° W.; position to start from S.S.W. three-fourths W. true, distant 155 miles.

Wednesday, June 23.—Clear weather, smooth sea, light breezes from W. by S.; steaming only; lat., noon, 51° 50′ N., long. 32° 48′; at P.M. made two sail ahead that soon prove to be the *Valorous* and *Gorgon;* the *Valorous* lowers boat and boards us; congratulations pass on escaping the fury of the gale; both these ships sustain some damage; the three of the Squadron on hand near the position; all heave to, to remain stationary till

Agamemnon arrives; at midnight it is foggy, neither the *Valorous* nor *Gorgon* to be seen.

Thursday, June 24.—The day commences foggy; ship heading W.N.W.; we drift rapidly to north at rate of two miles per hour per log, yet at noon find we have experienced a powerful south current of 35 miles. Nothing in sight at noon, 51° 15′, long. 23° 3′. After obtaining the latitude the *Niagara* squared yards, made sail, run before the wind, in order to recover lost latitude. At 7½, it commences to blow and increases to a gale; furled mizen-topsail and foresail; put ship under snug sail; the wind at N.W. Nothing seen of any of the Squadron all this day.

Friday, 25th, A. M.—Strong gales, clear weather; ship under close-reefed fore and main-topsails and storm fore-staysail; considerable sea. Noon, no observation; lat. by account, 52° 3′, long. by account, 33° 18′. At noon squared away to run north to make up lost northing, and also to search after the absent Squadron; wind S.S.W.; clear. At 7. P.M. blowing strong; close-reefed the topsails, furled foresail and mizen-topsail, wore ship and stood S.W.

Friday, June 25.—Strong gales, but clear, some sea—noon, lat. 52° 3′, long. 33° 18′ W.; at 1 P.M. saw 3 sail ahead. Soon discovered they were the Telegraph Squadron. Furled all sail and stood for them. The weather became beautiful. All the Squadron have boats down, visiting each other's ships. The Captain of the *Agamemnon* informs us experienced same gale we did—suffered heavy damages—the entire coil of Telegraph Cable had shifted; his ship was in a very critical situation, and he must re-coil a portion of the Cable before it could be ready for splicing, and could not be ready before 24 hours. Day ends very fine in every respect.

Saturday, June 26.—Calm, beautiful weather; Squadron close together; at 8 A.M. a telegraphic message from *Agamemnon* says: "Will be ready to splice at 9 o'clock." Preparations are immediately made; hawsers sent on board of her, as also the end of the Telegraph Cable. Everything auspicious and favorable—lat. 52° 2′, long. 33° 18′ W. We are 15 days out at noon. At 1 o'clock we commence paying out Cable. At 3 o'clock we had 200 fathoms out; it is calm; the hawser is cast off from the *Agamemnon*, and the *Niagara* commences to pay out in earnest, as does the *Agamemnon*. At 3 45 just three miles of Cable had been paid out, the pressure 2,360 pounds, when, at that instant, as the Cable was coming out of the circle, the Cable got out of one of the grooves of the wheel and into another groove, and, before it could be liberated, it parted, by being cut by a tar-scraper attached to the wheel. Thus the three miles from our ship was lost, and probably the same amount from the *Agamemnon*, as no doubt she did not make any attempt to save it. The ships then again neared each other;

spliced again at 5 P.M.; took in hawsers and commenced paying out under the most favorable prospects; sea smooth as a mill-pond, and weather calm. The sight is beautiful. The *Gorgon* heads the *Niagara*—the *Valorous* heads the *Agamemnon*. We follow the N.W. ¾ N.—the *Agamemnon* a S.E. ¾ S. course; our rate of speed, three miles per hour. At 8 P.M., 12 miles and 270 fathoms have run out most beautifully, the pressure being very uniform at 2,400 pounds. The same continues till midnight, when this day ends.

Sunday, June 27.—The sea continues smooth; wind light, from the westward; course and rate of sailing, the same. At 1 A.M. the electricians report the continuity has ceased. Every person who hears this report is struck with consternation, for everything seemed progressing so very fine and satisfactory; the dynamometer indicating 2,400 pounds; the rotometer 35 miles, 270 fathoms, as being out. Although this discovery is reported at 1 o'clock, A.M., the Cable is continued to be paid out slowly, and the ship's speed is reduced, until 5½ A.M. The Cable then out is 43 miles, 280 fathoms. The dynamometer rises from 3,000 to 3,200 and 3,400 pounds; ship's progress is slowed down to one mile per hour. The engine is set in motion to haul in Cable, and that task is commenced at 5½ o'clock. The Cable comes in slowly, the pressure increases to 4,300 pounds, at which point the Cable snaps, and all except about three-fourths of a mile hanging out is lost. What little we succeed in hauling in is very much snarled and badly kinked. The very long-jawed wire is nearly separated from the inside covering of gutta percha, and is totally unfit for any service. The ship is now put head about towards the point of starting, in order to re-connect the Cable the third time with the *Agamemnon*. The opinion prevails on board our ship, founded on some trials and scientific principles, that the cause of failure of continuity must have occurred close to or on board the *Agamemnon*. All confidence is given that actual trial proves the break did not occur on our side of the splice, but that it was between the splice and the *Agamemnon*. We were at this time 84 miles apart, and the splice was supposed to be midway. At 10½ A.M.—it is calculated we are on the precise spot from whence we started to lay the Cable. The *Agamemnon* is not here, but the *Gorgon* is. At 11 A.M. lowered a boat, and boarded the ship *Alice Munroe*, fourteen days from Liverpool, bound for Boston. Noon, light N.W. wind and a smooth sea—lat. 51° 57′ north, long. 32° 46′ west. Afternoon, calm, sea smooth, banked up fires. From midnight till noon fine weather.

Monday, June 28.—At 1 P.M. discovered three sail; soon made them out to be the Telegraph Squadron. At 4½ P.M. all three sent boats to the *Niagara*. At 6 P.M. sent hawser and Cable to *Agamemnon;* paid out 210 fathoms Cable; indicator standing at 582–168, and previous to paying out any, it was 582–100; wind quite light, at W.N.W.; all sail is furled on board both ships —the *Valorous* in position ahead of the *Agamemnon*, and the *Gorgon* ahead of the *Niagara;* the *Valorous* and *Agamemnon* heading S.E. ½ S.—the *Gorgon* and *Niagara* N.W. ¾ N.; our supposed depth of water, 1,670 fathoms, at 6 13 P.M.; we start for our respective destinations at 8 2 P.M. precisely, there is just three nautical miles of Cable paid out; at this time and quantity out the dynamometer first shows an indication of pressure of 2,100 pounds; up to midnight the speed of ship is 3¾ of a mile per hour—(by log hove every fifteen minutes)—while the Cable speed is 5¼ miles per hour; the pressure is very regular and uniform at 2,200 pounds; ends beautiful—the *Gorgon* close to us.

Tuesday, June 29.—Commences beautiful weather, sea perfectly smooth, wind light, at W.S.W.; course, N.W. ¾ N.; speed, 4 miles per hour, Cable going out finely. At 4 A. M. the *Niagara* has paid out 48 nautical miles; at 8 A. M., 68 nautical miles and 200 fathoms. The actual distance run from point of starting is 46 miles; pressure, 2,200 pounds; from 8 o'clock to noon, rate of sailing is 4 knots. At noon, the amount of Cable out is 89 miles, 360 fathoms; pressure, 2,200 pounds; fine weather, sea smooth; barometer 30°38′; air, 56°; water, 58°; lat. 54° 4′; long. 35° 2′. The latter part of this 24 hours is pleasant, wind at W.S.W.—speed of ship 4 knots. At 4 o'clock *one hundred and eleven miles* and 600 fathoms (nautical miles) have been paid out. The *Niagara* has scarcely any motion, and Cable runs out beautifully. At 9 hours and 10 minutes the electricians from their station give the fatal word, "*The continuity has stopped.*" 140 miles, 570 fathoms of Cable out up to this hour. From this time till 27 minutes after 12 midnight, the speed of the vessel was as slow as possible, and the Cable paid out as little as it could be. It had previously been understood, in the event of the failure of the continuity, six hours should elapse before paying-out should cease, or Cable suffered to part. Meanwhile, rockets were sent up, which the *Gorgon* answered, and she sent a boat to us. So ends this day. N. B.—The actual distance run since leaving the *Agamemnon* is 107 miles, up to midnight.

Wednesday, June 30.—The *Niagara* is riding wholly by the Telegraph Cable; speed of ship stopped; no motion to engine, and no paying out of Cable; pressure is applied in order to have Cable part, for it is now useless, but all to no purpose; 4,900 pounds are applied, and the weight of several persons upon the breaks; at 12 31 A.M. the Cable parts out from the stern; an attempt is made to haul it in, but in vain; the axe had to be applied, and 144 miles, 800 fathoms Cable, is lost on this trial. Thus ended trial No. 3. The Captain of the *Gorgon* now returns to his vessel, while Captain Hudson issues his orders to up helm and put on steam, and our ship is under full headway, bound for Cork harbor, Ireland, and at 8 A.M. set topsails. Afternoon, stiff breezes; driving more ahead; furled all sail. From this time till midnight heavy rain. Rate of sailing 8½ knots per hour.

Thursday, July 1.—Commences thick, rainy weather at 4 A.M.; set double-reefed topsails and foresail; speed 9½ miles per hour; course E. by S. one-half S.; wind S.W. by W., lat. 52° 26′; long. 29° 40′. Afternoon, overcast and squally; considerable rain at times; midnight rainy.

Friday, July 2.—Misty, rainy weather; moderate winds from W.N.W.; course E. by S.; sailed by log 258 miles; lat. 51° 55′; long. 23° 2′.

Saturday, July 3.—Fine, steady breezes; sea smooth; all sail set and full steam; speed 10 knots; sailed from noon to noon 230 miles; course E. by S., half S.; lat. 51° 15′ N. long. 17° 27′ W., 23 days from Plymouth; ends the same.

Sunday, July 4.—Fine breezes from W.N.W. throughout the day; lat 51° 17′ N.; long. 11° 54′ W.; at 6 P.M. made Cape Clear Light, distant 10 miles.

Monday, July 5.—Stiff breezes; ship lying by for pilot; at 2 A.M. took pilot on board; at 5 A.M. came to anchor, abreast of Queenstown, Cork harbor.

In laying the Cable, every possible care and attention

was given to the effort, and every preventive was used to guard against accident or misfortune. Even the captain and first lieutenant of the *Niagara* stood watch during the process of laying, day and night. The officer of the deck gave his constant attention to the log and courses, and the log was heaved every fifteen minutes. In the Cable-circle, twenty men were stationed. Ten men were at the paying-out machine, while another gang was stationed on the platform leading from the circle to the machine. The engineer of the Company was constantly at his post, or was relieved by the chief engineer of the *Niagara*. There were also two other engineers detailed to assist. Then there was a master's mate stationed at the brake; also two gentlemen connected with the Company, and the general business manager, all standing watch and regularly relieved; while one-half of the electricians were constantly on duty; in which department alone there were eight persons. The whole number of persons on board the *Niagara* in pay of the Company was twenty-two.

The scene at night was beautiful. Scarcely a word was spoken; silence was commanded, and no conversation allowed. Nothing was heard but the strange rattling of the machine as the Cable was running out. This music was singular, without variation. The lights about deck and in the quarter-deck circle added to the singularity of the spectacle; and those who were on board the ship describe the state of anxious suspense in which all were held as exceedingly impressive. The news of the successive disasters to the wire appeared to strike as though a personal hope had been extinguished.

The mode adopted by the ships in splicing was as follows:—The *Niagara* and *Agamemnon* made fast to each other—stern to stern—by a hawser. They kept 200 fathoms apart until two miles of Cable had been paid out (sufficient to reach bottom), then cast off and pursued separate courses at a rate generally of *five miles per hour*, while the ships' progress would vary from 3½ to 4½ miles per hour. The highest pressure at any time was 4,500 pounds, the lowest 1,900.

One remarkable circumstance attended the laying of the Cable. Every time the ships were prepared to splice, the weather was exceedingly fine, and the sea smooth, and so continued until the breaks occurred. Such was the case in every instance until the ships met again. In fact the only favorable weather was while the ships were engaged with the Cable.

The cruise of the *Agamemnon* was described as follows by an English Correspondent:—

"As we approached the place of meeting, the *Valorous* hove in sight at noon; in the afternoon the *Niagara* came in from the north, and in the evening the *Gorgon* from the south; and then, almost for the first time since starting, the squadron was reunited near the spot where the great work was to commence. The rendezvous actually agreed upon was 52° 2′ north lat., 33° 18′ west long., but the place where the vessels met was in 51° 54′ lat., 32° 33′ long., or about 30 miles more toward the English coast than had been agreed upon. On the evening of Friday, June 25, the four vessels lay together, side by side, and there was such a stillness in the sea and air (as would have seemed remarkable in an inland lake) on the Atlantic; and after what we had all so lately witnessed, it seemed almost unnatural. We have said how, during the awful rolls which the *Agamemnon* made on the 20th and 21st, the upper part of the main coil shifted, and became a mere shapeless tangled mass, with which it seemed impossible to deal in any conceivable way. For the first 24 hours the labor seemed hopeless, for so dense was the tangle, that an hour's hard work would sometimes scarcely clear half a mile. By-and-by, however, it began to mend, the efforts were redoubled, and late on Friday night 140 miles had been got out, and the remainder was found to be clear enough to commence work with.

"On the morning of Saturday, June 26, all the preparations were completed for making the splice, and commencing the great undertaking. The end of the *Niagara's* Cable was sent on board the *Agamemnon*, the splice was made, a bent sixpence put into it for luck, and at 2 50, Greenwich time, it was slowly lowered over the side and disappeared for ever. The weather was cold and foggy, with a stiff breeze and dismal sort of sleet, and as there was no cheering or manifestation of enthusiasm of any kind, the whole ceremony had a most funereal effect, and seemed as solemn as if we were burying a marine or some other mortuary task of the kind, equally cheerful and enlivening. It is needless making a long story longer, so we may state at once, that when each ship had paid out three miles or so, and were getting well apart, the Cable broke on board the *Niagara*, owing to its overriding and getting off the pulley leading on the machine. The break was, of course, known instantly; both vessels put about and returned; a fresh splice was made, and again lowered over at 7½ o'clock. According to arrangement, 150 fathoms were veered out from each ship, and then all stood away on their course, at first at two miles an hour, and afterwards at four. Everything then went well—the machine working beautifully, at 32 revolutions per minute—the screw at 26—the Cable running out easily at five and five and a half miles an hour, the ship going free. The greatest strain upon the dynamometer was 2,500 pounds, and this was only for a few minutes—the average giving only 2,000 pounds and 2,100 pounds. At 12 at midnight twenty-one nautical miles had been payed out, and the angle of the Cable with the horizon had been reduced considerably. At 3½ o'clock 40 miles had gone, and nothing could be more perfect and regular than the working of everything, when, suddenly, at 3 40 A.M., on Sunday, the 27th, Professor Thompson came on deck, and reported a total break of continuity; that the cable, in fact, had parted, and, as was believed at the time, from the *Niagara.* The *Agamemnon* was instantly stopped, and the brakes applied to the machinery, in order that the Cable payed out might be severed from the mass in the hold, and so enable Professor Thompson to discover by electrical tests at about what distance from the ship the fracture had taken place. Unfortunately, however, there was a strong breeze on at the time, with rather a heavy swell, which told severely upon the Cable, and ere any means could be taken to ease entirely the motion on the ship, it parted, a few fathoms below the stern-wheel, the dynamometer indicating a strain of nearly 4,000 pounds. In another instant a gun and a blue light warned the *Valorous* of what had happened, and roused all on board the *Agamemnon* to a knowledge that the machinery was silent, and that the first part of the Atlantic Cable had been laid and lost effectually.

"The great length of Cable on board both ships allowed a large margin for such mishaps as these, and the arrangement made before leaving England was that the splices might be renewed, and the work recommenced, till each ship had lost 250 miles of wire, after which they were to discontinue their efforts and return to Queenstown for orders. Accordingly, after the breakage on Sunday morning, the ships' heads were put about, and for the fourth time the *Agamemnon* again began the weary work of beating up against the wind for the everlasting rendezvous which we seemed destined to be always seeking. It was hard work beating up against the wind; so hard, indeed, that it was not till the noon of Monday, the 28th, that we rejoined the *Niagara;* and, while all were waiting with impatience for her explanation of how they broke the Cable, she electrified every one by running up the interrogatory, 'How did the Cable part?' This was astounding. As soon as the boats could be lowered, Mr. CYRUS FIELD, with the electricians from the *Niagara*, came on board, and a comparison of logs showed the painful and mysterious fact that, *at the same second of time*, each vessel discovered that a total fracture had taken place at a distance of certainly not less than ten miles from each ship—in fact, as well as can be judged, at the bottom of the ocean. That of all the many mishaps connected with the Atlantic Telegraph this is the worst and most disheartening is certain, since it proves that, after all that human skill and science can effect to lay the wire down with safety has been accomplished, there may be some fatal obstacles to success at the bottom of the ocean which can never be guarded against, for even the nature of the peril must always remain as secret and unknown as the depths in which it is to be encountered.

"No time was lost, after the vessels rejoined, in making

the third and last splice, which was lowered over into 2,000 fathoms water at 7 o'clock by ship's time the same night. The Cable, as before, paid out beautifully, and nothing could have been more regular and more easy than the working of every part of the apparatus. At first the ship's speed was only 2 knots, the cable going 3 and 3½, with a strain of 1,500 pounds, the horizontal angle averaging as low as 17, and the vertical about 16. By and by, however, the speed was increased to 4 knots, the cable going 5, at a strain of 2,000 pounds, and an angle of from 12 to 14. At this rate it was kept, with trifling variations, throughout almost the whole of Monday night, and neither Mr. Bright, Mr. Canning, nor Mr. Clifford ever quited the machines for an instant. Towards the middle of the night, while the rate of the ship continued the same, the speed at which the Cable payed out slackened nearly a knot an hour, while the dynanometer indicated as low as 1,300 pounds. This change could only be accounted for on the supposition that the water had shallowed to a considerable extent, and that the vessel was in fact passing over some submarine Ben Nevis or Skiddaw.

"After an interval of about an hour, the strain and rate of progress of the Cable again increased, while the increase of the vertical angle seemed to indicate that the wire was sinking down the side of a declivity. Beyond this there was no variation throughout Monday night, or, indeed, through Tuesday. The upper deck coil, which had weighed so heavily upon the ship, and still more heavily upon the minds of all during the past storms, was fast disappearing, and by 12 midday on Tuesday, the 29th, 76 miles had been paid out to something like 60 miles' progress of the ship. All seemed to promise most hopefully, and the only cause that warranted anxiety was that it was evident the upper deck coil would be finished by about 11 o'clock at night, when the men would have to pass in darkness along the great loop which formed the communication between that and the coil in the main hold. This was most unfortunate, but the operation had been successfully performed in daylight during the experimental trip in the Bay of Biscay, and every precaution was now taken that no accident should occur. At 9 o'clock by the ship's time, when 148 miles had been paid out, and about 112 miles' distance from the rendezvous accomplished, when the last flake but one of the upper deck coil came in turn to be used, in order to make it easier passing to the main coil, the revolutions of the screw were reduced gradually by two revolutions at a time from 30 to 20, while the Paying-out Machine went slowly from 36 to 22. At this rate, the vessel going three knots and the Cable three and a half, the operation was continued with perfect regularity, the dynamometer indicating a strain of 2,100 pounds. Suddenly, without an instant's warning, or the occurrence of any single incident that could account for it, the Cable parted. The gun that again told the *Valorous* of this fatal mishap brought all on board the *Agamemnon* rushing to the deck, for none could believe the rumor that had spread like wildfire about the ship. But there stood the machinery, silent and motionless, while the fractured end of the wire hung over the stern wheel, swinging loosely to and fro. It seemed almost impossible to realise the fact that an accident so instantaneous and irremediable should have occurred, and of course a variety of ingenious suggestions were instantly afloat, showing most satisfactorily how the Cable must and ought to have broke."

The amount of Cable lost from this ship was about one hundred and fifty miles; making an aggregate loss of about *three hundred miles of Cable*, during the progress of this Expedition.

The *Niagara* arrived at Queenstown on Monday, July 5,—one week in advance of the *Agamemnon*.

Captain HUDSON'S official report of the Expedition was as follows:

"UNITED STATES STEAM FRIGATE NIAGARA,
QUEENSTOWN, *Ireland*, *July* 8, 1858.

"SIR,—I am somewhat mortified and disappointed to report the arrival of the *Niagara* at this port on the 5th inst., after three unsuccessful attempts at laying down the Telegraphic Cable.

"My last dispatch, of the 10th ult., informed you that the squadron were off Plymouth harbor, bound to the appointed rendezvous for uniting and running out the Telegraphic Cable.

"During the first three or four days of our passage, we had calms and light variable winds; the following eight days almost continuous gales from the west to the south-west, and the greater part of the time heavy sea; when the weather again moderated, and our vessels, which had separated during the gales, met together at the rendezvous on the 25th; the *Agamemnon* having shifted about one hundred miles of the upper portion of the Cable on her main hold tier during the gale, which portion they were engaged in running to the gun-deck when we fell in with them.

"On the 26th (Saturday) we commenced our operations by securing the *Niagara* and *Agamemnon* together, stern to, with hawsers, splicing the Cable, and easing it down gradually with two hundred fathoms paid out from each ship; the hawser let go by signal, and the ships separated on their respective courses, at a rate of three-fourths of a mile the hour. When we had paid out two miles and forty fathoms, as shown by our indicator, the Cable, *being hauled in the wrong direction, through the excitement or carelessness of one of the men stationed by it*, caught and parted in the *Niagara's* machinery. A heavy fog and mist had set in soon after the ships separated.

We were fortunate enough, however, to get together again in a short time, splice, lower down the Cable, and separate from each other as before stated. The *Niagara's* speed at starting was short of one mile the hour, and gradually increased to two knots six fathoms up to 7 o'clock P.M., the Cable being paid out three and a half knots per hour; and from that hour till midnight, a uniform speed was maintained of three and a half miles the hour, and the Cable was paid out, as shown by the indicator, at four and a half miles the hour. Our machinery was working as well as we could desire, Cable running from the coils and going over it with ease and regularity, when to our great surprise, at 1¼ o'clock, A.M., on the 27th (Sunday), the electricians reported that there had been no signals from the *Agamemnon* for the last ten minutes. We kept going on slowly, as previously agreed upon, until 4 40 A.M. (in the meantime the electricians tested the Cable in the ship, and reported the continuity and insulations perfect), when the ship's headway was entirely stopped, and we commenced heaving in with the machinery. The Cable parted at 4 56 A.M., and we lost on this occasion, as measured by the indicator, 42 miles, 300 fathoms of Cable, and started for the rendezvous, where on Monday the 28th, the *Agamemnon* and *Niagara* were secured together, the splice made, lowered down, and the ships separated, as has been already described, at 7 30 P.M.

Our speed for the first hour was only three-quarters of a mile; second hour, 2½ miles; third hour, 3 miles; and the fourth hour, 3½ miles. From that time until 9 10 on Tuesday evening, the 29th (when we ceased to get signals from the *Agamemnon*, and the engines slowed down), the speed of the ship had been 4½ miles the hour, and the Cable paid out 5⅓ miles the hour, as shown by the indicator.

"The engines were stopped at 10 P.M., and the ship hung in a measure by the Cable until twenty minutes after midnight, when it parted, the indicator showing a loss on this occasion of 145 miles, 930 fathoms of the Cable. Our electricians again thoroughly tested all the Cable on board ship, and found the insulation and continuity all perfect, and there was but one opinion among those gentlemen, that the Cable parted at or near the *Agamemnon*, which we shall ascertain when she arrives at this port to fill up her coal.

"An arrangement had been made, when the ships separated on the 28th inst., that in the event of any accident to the Cable before either should have run *one hundred miles*, we were to return to the rendezvous, unite the Cable, and make another attempt to lay it out; if beyond that distance, the vessels were to proceed to Queenstown, fill up with coal, and again renew our efforts.

"In the *Niagara* we had all the coal that we required for laying down our part of the Cable. There were serious doubts, however, if we ran further, or any distance beyond the one named, whether the *Agamemnon*'s coal would hold out (without any expenditure in getting back to the rendezvous) and leave her enough to insure steaming back to Valentia Bay with the cable, in the event of no further casualty to it on the way there.

"Mr. EVERETT's machinery has paid out the Cable with apparent ease and uniformity of strain, and we find it admirably adapted to the work it has to perform in all its parts.

"Her Majesty's steamer *Gorgon*, which accompanied us, arrived here with the *Niagara*. We now await the arrival of the *Agamemnon* and the *Valorous*, when we hope to be off again for the rendezvous in seven or eight days, under more favorable auspices of weather than we experienced in the month of June.

"It affords me pleasure to report the continued good health of officers and crew.

"I have the honor to be, respectfully,

"Your obedient serv't,

"WM. L. HUDSON, Captain.

"Hon. I. TOUCEY, Secretary of the Navy, Washington, D. C."

The Electricians on board the *Niagara* made the following report:—

"At Sea, on Board U. S. Steam Frigate Niagara,
Electricians' Department, *July* 2, 1858.

To the Directors of the Atlantic Telegraph Company:

"Gentlemen—We beg to submit to you the following statement of the proceedings in our department during the expedition of the Telegraphic Squadron, from the sailing on the 10th to the 30th ult. On the passage out to the rendezvous we practised the staff of manipulating clerks in the working of the instruments until they were thoroughly proficient in the system. The instruments in circuit were a battery of 240 elements of copper and zinc, reversing key, magnetometer, ordinary galvanometer, Professor Thomson's marine galvanometer and four plug-switches, in conformity with your directions and the system of signalling authorized by your Board. On the 26th ult., at 1 58 P.M., Greenwich time, the *Niagara's* and *Agamemnon's* Cables having been united on board the latter ship, we commenced signalling through the Cable, (the whole of the *Niagara's* Cable being in circuit, with the exception of about 100 miles that had been cut out of circuit for the purpose of testing, there being faulty insulation in a part of it). We continued signalling through the Cable in the most perfect and satisfactory manner until 3 28 P.M., Greenwich time, when the Cable parted on board this ship. The telegraph ships then returned to the rendezvous, and Prof. Thomson came on board this ship to examine our record of signalling and testing with him. We arranged some new signals that did not materially alter the system authorized by you, but which increased our list of signals so as to meet some possible requirements, viz., signals to be used in case of the necessity of cutting and buoying the Cable, and also for any temporary stoppage that might occur.

"The two Cables were then spliced, and at 6 50 P.M., Greenwich time, we commenced passing signals to and from the *Agamemnon* until 3 32 A.M., Greenwich time, on the 27th, at which time we ceased receiving signals from the *Agamemnon*. After waiting fifteen minutes for the signals, and getting none, we informed Mr. EVERETT thereof. On applying all the different electrical tests in our power to the Cable, we found that there was 'dead earth' upon it, and informed the engineers of the fact. The Cable on board was then cut at about 100 miles from the bottom end and tested, and found perfect. Another cut was made, leaving about 196 miles in circuit on board, and this part of the Cable was found perfect. A third cut was made in the Cable, leaving only 20 miles in circuit on board, in addition to that paid out, the short end carried to our testing-room, and tested. All the Cable on board this ship was found to be perfect in all respects, but the tests applied to the Cable paid out showed that it was electrically broken, and that there was 'dead earth' upon it. The *Niagara* again returned to the rendezvous, and on Monday, June

28, having joined the *Agamemnon*, at Mr. EVERETT's request, Mr. DE SAUTY proceeded to the *Agamemnon* to confer with Professor THOMSON, and compare the records of the signals and tests. At 9 24 P.M., Greenwich time, the *Agamemnon's* and *Niagara's* Cables having been again united, we commenced signalling to and from the *Agamemnon* in the same perfect manner, until 11 44 P.M., Greenwich time, on the 29th, when, during the time we were receiving signals from the *Agamemnon*, they suddenly ceased, and on cutting the Cable, so as to have only about twenty miles in circuit in addition to that paid out, and testing it, we found 'dead earth' upon it. The *Niagara* then continued paying out slowly for a short time, and finally stoppered the Cable until 2 57 A.M. During this time we kept continually testing the Cable paid out, and every test showed that it was electrically broken a great distance from this ship.

"We cannot speak too highly of the system used for signalling, and can suggest no alteration or improvement, as we were not in the slightest degree embarrassed during the whole time that the Cable was being submerged.

"We are also very much pleased with the way in which the manipulating clerks, viz. Messrs. SMITH, GERHARDI, IRWIN, and LINDE, and Messrs MURRAY, MCFARLANE, and MORRISON attended to their duties. They did everything required of them most cheerfully and readily, and gave us every assistance in their power. We are, gentlemen, your most obedient servants,

C. V. DE SAUTY,
J. C. LAWS."

Chief Engineer EVERETT made report:—

"UNITED STATES STEAMER NIAGARA,
AT SEA, *June* 30, 1858.

To the Directors of the Atlantic Telegraph Company:

"GENTLEMEN—We beg to make the following report relating to the paying out of the Telegraph Cable from this ship. We sailed from Plymouth in company with the steamers *Agamemnon*, *Valorous* and *Gorgon*, on Thursday, June 10th inst., but did not meet the *Agamemnon* at the appointed rendezvous (latitude 52° 2′, longitude 33° 18′), until the 26th inst., as most of the voyage had been one continuous gale, and the vessels were unable to keep sight of each other. At 12 18, local time, the splice connecting the Cables of the two ships had been made, and we commenced paying out. At 1 45, the leading on the part of the Cable ran into the adjoining groove, and in the excitement of first starting, while attempting to put it into the proper groove, *it was thrown completely off the wheel, and was parted on the handle on the tar-scraper.* Two miles and forty fathoms of the Cable had been paid out.

"The splice was again made and we commenced paying

out at 5 20 P.M., the ship going slowly ahead, and the Cable running out at the rate of three and a half knots per hour until 7 o'clock, when the ship's speed was increased to three knots, and from this gradually to three and three-quarter knots, and the Cable was paid out from four and a half to five knots, the strain varying from 2,100 to 2,300 pounds. At 1 40 (27th) Mr. DE SAUTY, the electrician, reported that no signals had been received for the last quarter of an hour, and that from his tests he believed the Cable had parted at a considerable distance from the ship. The ship's speed was reduced as much as possible, in order to pay out the least amount of Cable practicable, while the electricians made further experiments. At 4 50 A.M., the electricians having given an unqualified opinion that the Cable was parted, we decided to attempt hauling in. The engines were connected, and about one hundred fathoms recovered, when the Cable parted near the surface of the water. The wind was fresh, with considerable sea. Forty-two miles and three hundred fathoms had been paid out, and the running of the Cable from the coils and the mechanical arrangements for paying out had been perfectly satisfactory.

"On the 28th, soon after midday, the ships again met at the rendezvous, when Mr. EVERETT visited the *Agamemnon* to confer with the engineers, and ascertained that the Cable had not been broken on board that ship, but that they had supposed it had been broken on board the *Niagara.* It is conclusive that the Cable must have parted *some distance from either ship*, but from what cause, or the precise place, we have no means of ascertaining.

"At 6 7 P.M. the splice was again made and lowered, the ship moving ahead slowly, and we payed out the Cable as before, until 8 o'clock, when the speed was increased to three knots, and further increased to four knots by midnight.

"At 12 o'clock M. (29th), by observation, the ship had run 67 miles, and we had paid out 89 miles and 360 fathoms of Cable. During the past twelve hours, the speed of the ship had averaged about 4½ knots, and 5½ knots of Cable had been paid out per hour. Nearly the same rate of speed of ship and Cable as before was maintained until 6 18 P.M., when signals were again reported to have failed by the electricians. The ship's speed was reduced, and the Cable paid out very slowly. At 11 o'clock, the electricians addressed us a note, and we determined to stop paying out and to let the ship ride by the Cable until it parted. Although the wind was quite fresh, *the Cable held the ship for one hour and forty minutes before breaking, and notwithstanding a strain of four tons.* By soundings on chart the water was 1,650 fathoms.

"The ship had run on her course 109 miles, and 142

miles, 280 fathoms of the Cable had been paid out, or about thirty per cent. more Cable than the distance run; but an allowance of ten miles at least must be made for the excess of Cable paid out immediately after the splice was made, which will reduce the per centage of loss to about twenty-one per cent. There had been at no time a strain of a ton upon the Cable since the splice was last made, and the angle at which it was running out varied from twelve to nineteen degrees with the horizon. The paying-out machinery worked perfectly, and we have not had the slightest difficulty in any department; and up to the time of the failure in the receipt of signals we had the utmost confidence in the successful termination of the enterprise.

"There is now remaining in the ship 1,000 miles of cable, or about thirty per cent. excess over the distance to be run, and should you think proper to renew the attempt, we feel confident there is sufficient Cable now in the ship to meet the requirements, and are ready to return so soon as the ship has obtained the necessary supply of coal.

W. E. Everett
W. H. Woodhouse."

The announcement of the departure of this Expedition had revived the anxiety with which every step of the enterprise was regarded. Tidings from the fleet were awaited on this side of the Atlantic in painful suspense.

Meanwhile, a stormier June than had been known on the Atlantic for many years, inspired fears for the result. Days passed away, and still no news came. Weeks fled, and yet no tidings. Finally, on Monday, July 13, when upwards of a month had passed, a brief dispatch from Newfoundland brought the first word from the fleet. The small steamer *Blue Jacket*, from Liverpool, reported that she passed a "large and a small steamer, both British," in a latitude near the place of rendezvous for the *Niagara* and *Agamemnon;* and that "in the evening," she "observed a large steamer bearing down upon the others." This indefinite information aggravated the condition of suspense. But a day or two afterwards came the *Alice Munroe*, a sailing packet, at Boston, with fuller particulars; and close upon the track of the *Alice Munroe*, a European steamer, bringing detailed accounts of the second failure.

The events which resulted in the final success, will form the subject of a fresh chapter.

CHAPTER VII.

THE THIRD AND SUCCESSFUL ATTEMPT.

THE *Agamemnon* and *Gorgon* having arrived at Queenstown on the 12th July, the news of the second great disaster was too fully confirmed. The Expedition had not only ended, but three hundred miles of the Cable had been lost in mid-ocean, during this second trial. Doubts and fears again gathered thickly around the enterprise. Prophets of evil, flattering themselves that their apprehensions were firmly grounded, wisely shook their heads, and indulged in self-complacent observations upon their forecast and prudence. Worse than all, the credit of the Company fell rapidly. On the 6th of July, on the receipt of the news of the return of the *Niagara* in London, the £1000 shares of the Company receded from £600, at which they had been nominally quoted, to £200. The closing rates on that day ranged from £200 to £400. That want of confidence, which has wrecked so many enterprises and blasted so many hopes, returned in full vigor upon the heads of the unfortunate projectors of this great undertaking. It is not necessary to add that the reception of the untoward intelligence in the United States chilled the popular enthusiasm, and created doubts in the minds of the most sanguine. The Atlantic Cable was virtually consigned to the long catalogue of impracticable fallacies.

Still, the noble-hearted and self-denying Directors did not falter. A special meeting of the Company was called, to meet in London early in July, in which a resolution to put forth a new effort was agreed to unanimously. It was found that the wise provision of an extra length of the Cable had relieved the Company from the necessity of further delay for the manufacture of a new supply. More than the necessary quantity of wire to compass the ocean yet remained unharmed on board the *Niagara* and *Agamemnon*, notwithstanding the heavy losses incurred during the second trip. The weather, in the latter part of the summer, promised more favorably than the remarkably erratic course of the winds and waves in June. The officers and men employed in the fleet were full of enthusiasm for the work. These were all valid reasons for renewing the effort. The summer waned, and no time was to be lost. Without wasting words in elaborate discussion, the Directors took their course, the Expedition was again ordered to sea, and the enterprise went on.

The last ships of the Telegraphic Squadron arrived at

Queenstown on the 12th of July, and on Saturday, July 17th, the entire fleet was again under way, bound to the mid-ocean rendezvous.

At noon on Thursday, August 5, the city of New York was electrified by the announcement of the arrival of the *Niagara* and her tender, the *Gorgon*, at Trinity Bay, Newfoundland; with the astounding tidings that the Atlantic Cable was safely laid, and already in perfect working order.

There is a homely proverb which says of unexpected intelligence, that "it is too good to be true." The adage was thoroughly verified in this instance. Men were generally incredulous. The sanguine shook their heads in doubt. That the news of a complete success, after so many disasters, should have come with the suddenness of a flash along the wire itself, was an event of no common moment. The fact failed to obtain credence for some hours after its announcement. Then came confirmatory despatches from Trinity Bay. Doubts were set aside, and the whole country broke out in uproarious rejoicings. The confirmation of the news being immediately telegraphed from New York to all parts of the United States, the event was made the occasion of *impromptu* celebrations without number. Bells were rung, bonfires blazed, business ceased, illuminations sprang up, the Press became jubilant. Rarely has there been heard so universal a shout of joy.

The first telegram announcing the success of the enterprise, was sent from Trinity Bay by Mr. Cyrus W. Field, directed to the Press of New York. It was as follows:—

"Trinity Bay, Thursday, Aug. 5, 1858.

"*To the Associated Press of New York:*

"The *Niagara* and *Gorgon* arrived at Trinity Bay yesterday, and the Atlantic Cable, the working of which is perfect, is being landed to-day.

"The Atlantic Telegraph Fleet sailed from Queenstown on Saturday, July 17, met at mid-ocean on Wednesday, the 28th, made the splice at 1 P.M. on Thursday, the 29th, and then separated, the *Agamemnon* and *Valorous* bound to Valentia, Ireland, and the *Niagara* and *Gorgon* for this place, where they arrived yesterday, and this morning the end of the Cable will be landed. It is sixteen hundred and ninety-eight nautical or nineteen hundred and fifty statute miles from the Telegraph house, at the head of Valentia harbor, to the Telegraph house, Bay of Bulls, Trinity Bay, and for more than two-thirds of this distance the water is over two miles in depth. The Cable has been paid out from the *Agamemnon* at about the same speed as from the *Niagara*. The electrical signals sent and received through the whole Cable are perfect. The machinery for paying out the Cable worked in the most satisfactory manner, and was not stopped for a single moment from the time the splice was made until we arrived here.

"Captain Hudson, Messrs. Everett and Woodhouse, the Engineers, the Electricians and officers of the ships, and in fact every man on board the Telegraph Fleet, has exerted himself to the utmost to make the expedition successful; and by the blessing of Divine Providence it has succeeded.

"After the end of the Cable is landed and connected with the land line of telegraph, and the *Niagara* has discharged some cargo belonging to the Telegraph Company, she will go to St. John's for coals, and then proceed at once to New York.

"CYRUS W. FIELD."

Subsequent dispatches, exchanged between Mr. FIELD and the President of the United States, the Mayor of the City of New York, and others, added to the fever of the popular excitement. We present these dispatches, as a valuable portion of the history of the Telegraphic enterprise:—

DISPATCH TO THE PRESIDENT OF THE UNITED STATES.

To the President of the United States, Washington:

DEAR SIR: The Atlantic Telegraph Cable on board the United States frigate *Niagara*, and H.B.M. steamer *Agamemnon*, was joined in mid-ocean, July 29, and has been successfully laid; and as soon as the two ends are connected with the land lines, Queen

VICTORIA will send a Message to you, and the Cable will be kept free until after your Reply has been transmitted.

With great respect,

I remain your obedient servant,

CYRUS W. FIELD.

THE PRESIDENT'S REPLY.

BEDFORD, Penn., Friday, Aug. 6, 1858.

To Cyrus W. Field, Trinity Bay:

MY DEAR SIR: I congratulate you with all my heart on the success of the great enterprise with which your name is so honorably connected. Under the blessing of Divine Providence I trust it may prove instrumental in promoting perpetual peace and friendship between the kindred nations.

I have not yet received the Queen's dispatch.

Yours very respectfully,

JAMES BUCHANAN.

SECOND DISPATCH TO THE PRESIDENT.

TRINITY BAY, Saturday, Aug. 7, 1858.

His Excellency James Buchanan, President of the United States, Bedford Springs:

Your telegraphic dispatch has been received. We landed here in a wilderness, and until the telegraph instruments are all perfectly adjusted, no message can be recorded over the Cable. You shall have the earliest information, but some days may elapse before all is effected. The first message from Europe shall be from the Queen to yourself, and the first from America to England, your reply.

With great respect,

Very truly your friend,

CYRUS W. FIELD.

DISPATCH TO THE MAYOR OF NEW YORK.

TRINITY BAY, *Thursday, Aug.* 5, 1858.

Mayor of New York:

SIR: The Atlantic Telegraph Cable has been successfully laid.

C. W. FIELD.

THE MAYOR'S REPLY.

MAYOR'S OFFICE,
NEW YORK, *Friday, Aug.* 6, 1858.

Cyrus W. Field, Esq., Trinity Bay:

SIR: Your dispatch has been received. I congratulate you myself, and for the people of this City, on the success of the great work of uniting the old and the new worlds by the Electric Telegraph. Science, and skill, and perseverance, have finally triumphed.

DANIEL F. TIEMANN.

A journal of the third and last voyage of the *Niagara*, carefully posted from day to day by Mr. FIELD, and published by his permission, embodies a complete history of the final triumph. It is as follows:—

"Saturday, July 17.—This morning the Telegraph Fleet sailed from Queenstown, Ireland, as follows: The *Valorous* and *Gorgon* at 11 A.M.; the *Niagara* at 7½ P.M., and the *Agamemnon* a few hours later. All the steamers are to use coal as little as possible in getting to the rendezvous. Up to 5 P.M. clear weather and blue sky; from 5 to 9 P.M. overcast, threatening weather and drizzling rain; from 9 to 12 P.M., overcast, hazy and squally.

Sunday, 18th.—The *Niagara* passed Cape Clear in the morning. Wind varying from W. by W.N.W. Heavy atmosphere, cloudy and squally.

Monday, 19th.—Wind varying from W. to N.W. Hazy atmosphere, cloudy and rainy.

Tuesday, 20th.—Wind from N.W. to N. Hazy atmosphere, cloudy and squally.

Wednesday, 21st.—Wind N.W. with slight variation to the eastward, and cloudy.

Thursday, 22nd.—Wind N.W. by W. Blue sky and cloudy.

Friday, 23rd.—Wind from W. by S. to W.S.W. and cloudy and hazy atmosphere and rain. The *Niagara* arrived at rendezvous, lat. 52° 5′, long. 32° 40′, at 8 30 P.M.

Saturday, 24th.—Wind W.N.W. and hazy atmosphere, cloudy and squally.

Sunday, 25th.—*Valorous* arrived at 4 P.M. Calm, hazy atmosphere and cloudy.

Monday, 26th.—Calm, hazy atmosphere, cloudy. Capt. OLDHAM of the *Valorous* came on board of the *Niagara.*

Tuesday, 27th.—Calm, hazy atmosphere. *Gorgon* arrived at 5 P.M.

Wednesday, 28th.—Slight wind N.N.W., blue sky and hazy atmosphere. *Agamemnon* arrived at 5 P.M.

Thursday, 29th.—Lat. 52° 9′ north, long. 32° 27′

west. Telegraph Fleet all in sight; sea smooth; light wind from S.E. to S.S.E., cloudy. Splice made at 1 P.M. Signals through the whole length of the Cable on board both ships perfect. Depth of water 1500 fathoms; distance to the entrance of Valentia harbor 813 nautical miles, and from there to the telegraph house the shore end of the Cable is laid. Distance to the entrance of Trinity Bay, Newfoundland, 822 nautical miles, and from there to the telegraph house at the head of the bay of Bull's Arm, 60 miles, making in all 882 nautical miles. The *Niagara* has 69 miles further to run than the *Agamemnon.* The *Niagara* and *Agamemnon* have each 1100 nautical miles of cable on board, about the same quantity as last year. At 7 45 P.M. ship's time, or 10 5 P.M. Greenwich time, signals from the *Agamemnon* ceased, and the tests applied by the electricians showed that there was a want of continuity on the Cable, but the insulation was perfect. Kept on paying out from the Niagara very slowly, and constantly applying all kinds of electrical tests until 9 10 ship's time and 11 30 P.M. Greenwich time, when again commenced receiving perfect signals from the *Agamemnon.*

Friday, 30th.—Lat. 51° 50′ N. Long. 34° 49′ W., distance run by observation the last 23 hours 89 miles. Paid out 131 miles, 900 fathoms Cable, or a surplus of 42 miles, 900 fathoms, over the distance run by observation, equal to 48 per cent. Depth of water 1550 to 1975 fathoms;

wind from S.E. to S.W. Weather thick and rainy, with some sea; *Gorgon* in sight. At 5 30 A.M. finished the main deck coil, and commenced paying out from the berth deck. 723 miles from telegraph house at Bull's Arm, Trinity Bay. Friday 30, at 2 21 P.M. received signals from the *Agamemnon* that they had paid out 150 miles.

Saturday, 31st.—Lat. 51° 5′ north, long. 38° 14′ W; distance run from observation the last four hours, 137 miles; payed out 156 miles, 843 fathoms of Cable, or a surplus of 22 miles; 843 fathoms over the distance run by observation, equal to 17 per cent.; depth of water, 1657 to 2250 fathoms: wind moderate, S.W., and from 6 A.M. N.W. by N.; weather cloudy; little rain and some sea: *Gorgon* in sight. Total amount of Cable paid out 291 miles; 730 fathoms; total distance run by observation, 226 miles. Surplus Cable payed out over the distance run by observation, 65 miles, 730 fathoms, equal to 29 per cent.; 656 miles from the telegraph house at Trinity Bay; 11 4 P.M., payed out from the *Niagara* 300 miles of Cable; at 2 45 P.M. received signals from the *Agamemnon*, that they had payed out from her 300 miles of Cable; at 5 37 P.M., finished coil on the berth, and commenced paying out from the lower deck.

Sunday, Aug. 1.—Lat. 50° 32′ N. long. 41° 55′ W.; distance run by observation the last 24 hours, 145 miles; payed out 164 miles and 683 fathoms of Cable, or a surplus of 19 miles, 630 fathoms over the distance run by observation, equal to 14 per cent.; depth of water 1,950 to 2,424 fathoms; wind moderate and fresh from N.N.E to N.E.—weather cloudy, misty, and a heavy swell; *Gorgon* in sight. Total amount of Cable payed out 456 miles, 400 fathoms; do. do. run by observation, 371. Surplus Cable payed out over the distance run, 85 miles, 600 fathoms, equal to 23 per cent.; 511 miles from the telegraph house; at 3 05 P.M., finished paying out coil on the lower deck, and changed to the coil in the hold.

Monday, 2d.—Lat. 49° 52′ N. lon. 45° 48′ W.; distance run by observation the last 24 hours, 154 miles; payed out 177 miles, 15 fathoms of Cable, or a surplus of 23 miles, 100 fathoms over the distance run, equal to 15 per cent. Depth of water 1,600 to 2,385 fathoms. Wind N., weather cloudy; the *Niagara* getting light, and rolling very much; it was not considered safe to carry sail to steady ship, for in case of accident it might be necessary to stop the vessel as soon as possible. At 7 A.M., passed and signalled the Cunard steamer from Boston to Liverpool. Total amount of Cable payed out, 633 miles, 500 fathoms; run by observation, 525 miles; total surplus of Cable payed out over the distance run, 108 miles, 500 fathoms, or less than 21 per cent.; 257 miles from the telegraph house; 12 38 ship's time, 3 38 Greenwich time, imperfect insulation of Cable detected in sending and receiving signals from the *Agamemnon*, which continued until 5 40 A.M. ship's time, or 8 40 A.M., Greenwich time, when all was right again. The fault was found to be in the ward room, or in about 60 miles from the lower end, which was immediately cut out, and taken out of the circuit.

Tuesday, 3d.—Lat. 45° 17′ N. long. 49° 23′ W.; distance run, by observation, the last 24 hours, 147 miles; payed out 161 miles 61 fathoms of Cable, or a surplus of 14 miles 613 fathoms, over the distance run,—equal to 10 per cent. Depth of water 742 to 1,827 fathoms; wind N.N.W.; weather very pleasant; *Gorgon* in sight; total amount of Cable paid out, 796 miles, 300 fathoms. Run by observation 672 miles; surplus of Cable payed out, over the distance run, 123 miles, 300 fathoms—less than 19 per cent.; 200 miles from the telegraph house; at 8 26 A.M., finished paying out coil from the hold, and commenced paying out from the ward room coil; 305 miles of Cable on board at noon; at 11 15 ship's time, received signals from on board the *Agamemnon*, that they had paid out from her 780 miles of Cable; on the afternoon and evening, passed several icebergs; at 9 10 P.M., ship's time, received signals from the *Agamemnon* that she was in water of 200 fathoms; at 10 20 P.M., ship's time, *Niagara* in water of 200 fathoms, and informed the *Agamemnon* of the same.

Wednesday, 4th.—Lat. 48° 17′ N. long. 52° 43′ W.; distance run, by observation, 146 miles; payed out 154 miles, 160 fathoms of Cable, or a surplus of 8 miles—360 fathoms over the distance run—equal to 6 per cent.; depth of water less than 200 fathoms; weather beautiful, perfectly calm; *Gorgon* in sight. Total amount of Cable paid out, 949 miles, 660 fathoms; surplus of Cable paid out, over the distance run, 131 miles, 666 fathoms; 64 miles to telegraph house; received signals from the *Agamemnon*, at noon, that they had paid out from her 940 miles of Cable. Passed several icebergs this morning; made the land off the entrance to Trinity Bay, at 8 A.M.; entered Trinity Bay at 12 30 P.M., ship's time; stopped sending signals to the *Agamemnon*, for the purpose of making a splice. At 2 40 ship's time, commenced sending signals again to the *Agamemnon;* at 5 P.M. saw H.B.M. steamer *Porcupine*, coming to us; at 7 30 P.M., Captain OTTER of the *Porcupine*, who had been for the last six weeks surveying and sounding Trinity Bay, came on board the *Niagara*, to pilot us to anchorage near the telegraph house.

Thursday, 5th.—1 45 A.M.; *Niagara* anchored. Distance run since yesterday noon, 64 miles; amount of cable paid out, 66 miles, 353 fathoms—being a loss of less than 4 per cent. Total amount of cable paid out since the splice was made, 1,016 miles, 600 fathoms. Total amount of distance run, 882 miles; amount of Cable paid out over the distance run, 134 miles, 600 fathoms—being a surplus of 15 per cent. At 2 A.M.

went ashore in a small boat, and informed the persons in charge of the telegraph house, half a mile from the landing, that the Telegraph Fleet had arrived, and were ready to land the end of the cable; 2 45 A.M., received signal from the *Agamemnon* that they had paid out from her 1,101 miles of Cable. At 5 15 A.M. Telegraph Cable landed; at 6 the shore end of the Cable was carried into the telegraph house, and received a very strong current of electricity from the other side of the Atlantic. Capt. HUDSON, of the *Niagara*, then read prayers, and made some remarks. 1 P.M., H.M. steamer *Gorgon* fired a royal salute of 21 guns. All day discharging cargo belonging to Telegraph Co.; all day Friday receiving strong electric signals from the telegraph house at Valentia.

NOTE.—We landed here in the woods; until the telegraph instruments are all ready, and perfectly adjusted, communications cannot pass between the two continents, but the electric currents are received freely. You shall have the earliest intimation when all is ready, but it may be some days before everything is perfected. The first through message between Europe and America will be from the Queen of England to the President of the United States, and the second, his reply.

CYRUS W. FIELD.

A period of twelve days elapsed from the date of Mr. FIELD'S first dispatch until the reception of the

Queen's Message to the President of the United States. During this interval, daily dispatches were received in New York from the operators at Trinity Bay, giving positive assurances that the Cable was in perfect working order, and that signals were constantly passing between the British and American termini of the line. Still, these assurances aroused a new suspicion in the public mind. The Cable, men said, might have been safely laid, signals might be received hourly through its entire length, the operation of the electric current might remain unimpeded,—but could messages be sent through it? The question was one that no one answered; nor did the electricians in charge at Trinity Bay vouchsafe the explanation that would have set doubts at rest. Days passed, and yet the promised Message from Her Majesty did not appear. Intense anxiety began to prevail. The doubting part of the public renewed their prophecies of evil. Success was still far from positive certainty,—when suddenly, in the afternoon of Monday, August 16, the tidings reached New York from Newfoundland, that the Queen's Message was received. A few hours afterwards, a short paragraph, purporting to be the congratulations of Her Majesty to the President, came to us. A general feeling of disappointment at the remarkable brevity, not to say curtness, of this message, found vigorous expression. The President himself, who returned to the White House to receive the message, entertained doubts of its genuine character, but upon receiving assurances of its correctness, dispatched a reply. The following morning—Tuesday, August 17—brought the explanation of the matter. The message sent on Monday was but a small part of Her Majesty's communication: the wires had ceased to work when the introductory paragraph was dispatched, but the mistake was rectified, the message transmitted entire, and the anxiety of the public was allayed,—for the Cable was a working instrument—an accomplished fact.

The message of Her Majesty and the reply of the President, were as follows:

MESSAGE OF THE QUEEN.

To the President of the United States, Washington:

The Queen desires to congratulate the President upon the successful completion of this great international work, in which the Queen has taken the deepest interest.

The Queen is convinced that the President will join with her in fervently hoping that the Electric Cable which now connects Great Britain with the United States will prove an additional link between the nations whose friendship is founded upon their common interest and reciprocal esteem.

The Queen has much pleasure in thus communicating with the President, and renewing to him her wishes for the prosperity of the United States.

REPLY OF THE PRESIDENT.

WASHINGTON CITY, Aug. 16, 1858.

To Her Majesty VICTORIA, *Queen of Great Britain:*

The President cordially reciprocates the congratulations of Her Majesty, the Queen, on the success of the great international enterprise accomplished by the science, skill, and indomitable energy of the two countries. It is a triumph more glorious, because far more useful to mankind, than was ever won by conqueror on the field of battle.

May the Atlantic Telegraph, under the blessing of Heaven, prove to be a bond of perpetual peace and friendship between the kindred nations, and an instrument destined by Divine Providence to diffuse religion, civilization, liberty, and law throughout the world. In this view will not all nations of Christendom spontaneously unite in the declaration that it shall be for ever neutral, and that its communications shall be held sacred in passing to their places of destination, even in the midst of hostilities? (Signed)

JAMES BUCHANAN.

Immediately after the transmission of the President's reply, the Mayor of the City of New York sent the following message to the Lord Mayor of London:

MAYOR'S OFFICE, NEW YORK, Aug. 17, 1858.

To the Right Honorable Sir ROBERT WALKER CARDEN, M. P., *Lord Mayor of London:*

I congratulate your Lordship upon the successful laying of the Atlantic Cable, uniting the continents of Europe and America and the Cities of London and New York, the work of Great Britain and the United States, the triumph of science and energy over time and space, thus uniting more closely the bonds of peace and commercial prosperity, and introducing an era in the world's history pregnant with results beyond the conceptions of a finite mind. To God be all the praise.

DANIEL F. TIEMANN,
Mayor of New York City.

The Governor-General of Canada, on the same day, forwarded his congratulations to the Home Government, in the following dispatch:

The Honorable the Secretary of State for the Colonies, London, England:

The Governor-General of British North America presents his humble duty to the Queen, and respectfully congratulates Her Majesty on the completion of the Telegraphic communication between Great Britain and these Colonies.

EDMUND HEAD.

The frigate *Niagara*, having accomplished the task allotted to her, returned to the port of New York on the afternoon of Wednesday, August 18, after an absence of five months and nine days. On the 10th of August, before the departure of the *Niagara* from St. John's, a banquet was given by the authorities in honor of Mr. FIELD, who responded in a brief speech to a toast given in his honor.

In response to a complimentary address from the

Executive Council of Newfoundland, Mr. FIELD wrote the subjoined letter:

To the Honorable the Executive Council of Newfoundland:

MR. PRESIDENT AND HONORABLE GENTLEMEN: I thank you with all my heart for this cordial manifestation of your good will. There is, however, nothing new to me in the present tone of your feelings.

Upward of four years ago, when first I laid before the Legislature of this colony the plan of uniting the two continents by means of Telegraphic communication, I received your ready countenance, and in the Charter of Incorporation then passed was unfolded the whole view which has now arrived at its final accomplishment. The terms of that charter were liberal and encouraging; but had your councils been guided by a different spirit, the project would have been abandoned, and years perhaps might have passed without witnessing this happy union of the two worlds, with the beneficent consequences it is destined to diffuse.

The exclusive privileges conferred by the colony on the New York, Newfoundland, and London Telegraph Company, have been the subject of hostile criticism, and it is therefore with satisfaction I observe the approving terms in which you refer to them. Every enlightened country recognises a right of property in those who originate a work where science or skill or capital has been invested. This protection is necessary to draw out the efforts of men in new works of public utility, for who would sow if he could not reap? And while the individual has his reward, society is the gainer by his labors. In the exclusive privileges you have conferred on the Company I represent, the principle of copyright only is involved, and I think there can now be no doubt that your policy has conserved the interests of the colony; while I confidently trust the future may be productive of much benefit to your people from the great work which, from the beginning to the present time, has had your consistent and liberal support. I shall look with peculiar pleasure on the advantages you may derive from the proud position of this colony in the Telegraph connexion of the Old and New Worlds, and shall be ever ready to promote your views of advancement by all means in my power.

I am, honorable gentlemen,

Your faithful servant,

CYRUS W. FIELD.

In another communication, addressed to the Chamber of Commerce of St. John's, Mr. FIELD paid the following deserved tribute to gentlemen who have taken an active interest in the success of this enterprise:

I could not do justice to my own feelings, did I fail to acknowledge how much is owing to Capt. HUDSON and the officers of the *Niagara*, whose hearts were in the work, and whose toil was unceasing.

To Commander DAYMAN, of her Majesty's steamer *Gorgon*, for the soundings so accurately made by him the last year, and for the perfect manner in which he led the *Niagara* in the great "circle arc" while laying the Cable.

To Capt. OTTER, of her Majesty's steamer *Porcupine*, for the careful survey made by him in Trinity Bay, and for the admirable manner in which he piloted the *Niagara* at night to her anchorage.

To Mr. EVERETT, who has for months devoted his whole time to designing and perfecting the beautiful machinery that has so successfully paid out the Cable from the ships—machinery so perfect in every respect, that it was not for one moment stopped on board the *Niagara* until she reached her destination in Trinity Bay.

To Mr. Woodhouse, who superintended the coiling of the Cable, and zealously and ably co-operated with his brother engineer during the progress of paying out.

To the electricians, for their constant watchfulness.

To the men, for their almost ceaseless labor; and I feel confident that you will have a good report from the Commander, engineers, electricians, and others on board the *Agamemnon* and *Valorous*, the Irish portion of the fleet.

To the Directors of the Atlantic Telegraph Company, for the time they have devoted to the undertaking, without receiving any compensation for their services, and it must be a pleasure to many of you to know that the Director who has devoted more time than any other, was, for many years, a resident of this place, and well-known to all of you. I allude to Mr. Brooking, of London. To Mr. C. M. Lampson, a native of New England, but who has for the last twenty-seven years resided in London, who appreciated the great importance of this enterprise in both countries, and gave it most valuable aid, bringing his sound judgment and great business talent to the service of the Company.

To that distinguished American, Mr. George Peabody, and his most worthy partner, Mr. Morgan, who not only assisted it most liberally with their means, but to whom I could always go with confidence for advice.

I shall rejoice to find that the commercial interests of this colony, which you represent, may be largely benefited by the close bonds that will now be drawn by the agency of the Atlantic Telegraph between them, and the varied relations they hold throughout the world; and wishing you all every prosperity and happiness,

I am your very grateful friend,

Cyrus W. Field.

In closing this record of the progress of the Atlantic Telegraph, it would be unjust to pass unnoticed the public demonstrations of joy at the success of the undertaking. The whole Union rejoiced together. Every city, town, village, and hamlet in all parts of the country, sought to testify its sense of the importance of the work. The proclamation of a peace after an exterminating warfare, could not have produced a more general outburst of enthusiastic congratulation. It is no unimportant part of the history of the enterprise to specify some of these spontaneous demonstrations; a considerable space in the Appendix of this work is accordingly devoted to a general summary of the popular demonstrations which occurred in different parts of the United States, on the reception of tidings of complete success—demonstrations remarkable alike for their spontaneity and wild enthusiasm.

The Atlantic Telegraph is therefore a fact. A wonderful work has been accomplished within the compass of a few months. But two or three years elapsed from the inception of the enterprise to its triumphant accomplishment. England and America are placed within whispering distance of each other: a new link in the chain of destiny has been forged; the electric current binds two great nations together in bonds of amity: the world has made a gigantic stride in the path of progress:—the men who live in this day have reason to hope for the accomplishment of undertakings hitherto considered impossible, and a new era dates from the laying of a Cable in the Ocean.

CHAPTER VIII.

WORKING THE ATLANTIC TELEGRAPH—THE TERMINI OF THE LINE.

TIME must test the practical operation of the Atlantic Telegraph. Messages have passed over its entire length, the Cable is in working order, and the electricians express satisfaction at the operation of the current. But there is still a doubt whether a single small Cable will suffice for the accommodation of the business which must flow to this oceanic line. Another Cable, perhaps a number, will probably be found requisite to meet the requirements of Commerce. With new applications to meet the increasing demand, will come improvements suggested by experience. For the present, the initial enterprise accomplishes all that we can justly expect; and with the future lies the solution of the difficulties which have accompanied the beginning of the enterprise.

The primary source of the influence which was charged with the service of Atlantic Telegraphy, was a giant voltaic battery, of ten capacious cells, termed the

"Whitehouse Laminated or Perpetual Maintenance Battery," a title designating one peculiarity which especially fits it for employment. This battery is made upon the Smee principle, so far as the adoption of a quadrangular trough of gutta percha, wood strengthened outside, in which dilute acid is contained, the proportion of acid to water being one part in 15 or 16. There are grooves in the gutta percha, into which several metal plates slide in a vertical position. These plates are silver and zinc alternately, but they are not pairs of plates in an electrical sense. Each zinc plate rests firmly at the bottom on a long bar of zinc, which runs from end to end of the trough, and thus virtually unites the whole into one continuous extent of zinc, presenting not less than 2,000 square inches of excitable surface to the exciting liquid. Each silver plate hangs in a similar way from a metallic bar, which runs from end to end of the trough above, the whole of the silver being thus virtually united into one continuous surface of equal extent to the face of the zinc. The zinc does not reach so high as the upper longitudinal bar, and the silver does not hang down as low as the inferior longitudinal bar. The battery is thus composed of a single pair of laminated plates, although to the eye it seems to be made up of several pairs of plates. Nature has set the example of arranging extended surface into reduplicated folds, when it is required that such surface shall be packed away in a narrow space at the same time that a large acting area is preserved, in the laminated antennæ of the cockchafer. The antennæ, indeed, are the types of the Whitehouse battery. If any one of these reduplicated segments of either kind of metal is removed, the remaining portion continues its action steadily, the effect merely being the same that would be produced if a fragment of an ordinary pair of plates were temporarily cut away. The silver laminæ are of considerable thickness, and securely "platinated" all over—that is, platinum is thrown down upon their surfaces in a compact metallic form, and not merely in the black pulverulent state; consequently they are almost exempt from wear. Each zinc lamina is withdrawn as soon as its amalgamation is injuriously affected, or so soon as its own substance is mainly eaten away by the action of the chemical menstruum in which it is immersed, and a freshly amalgamated, or new zinc lamina, is inserted into its place. The capability of the piecemeal renewal of the consumptive element of the battery in this interpolatory and fragmentary way, is then the cause of its "perpetual maintaining" power. The intensity of a voltaic arrangement depends upon the number of its pairs of plates, or cells. If, in the experiment, the intensity of the electricity had been increased, without any alteration of quantity, merely by multiplying the number of the cells engaged, or by some analogous modification of instrumental agency, the body which resisted the current of the battery with such complete effect, would have been flashed through and burnt up, like the fragment of metal that had inferior powers of resistance.

The flashes of light and crackling sparks produced on making and breaking contact with the poles of this grand battery, are very undesirable phenomena in one particular. They are accompanied by a considerable waste of the metal of the pole. Each spark is really a considerable fragment of the metal absorbed into itself by the electrical agent, so to speak, and flown away with by it. To avoid this danger, an ingenious contrivance of the Electrician of the Company will be used. First, he arranged a set of twenty brass springs, something of the form and appearance of the keys of a musical instrument, in opposite pairs, so that a round horizontal bar, turning pivot-ways on its own centre, and flattened at the top, could lift by an edge either of the sets of ten springs, right or left as it was turned. This enabled the contact to be distributed through the entire length of the edge and breadth of the brass springs, and the course of the current to be reversed, accordingly as the right or left edge (the bar being worked by a crank-handle) was raised to the right or left set of springs—the right set, it will be understood, being the representatives of one pole of the battery, and the left set of the other pole. By this arrangement four-fifths of the sparks were destroyed, simply on account of the large surface of metal, through which the electrical current had to pass when contact was completed. Still there remained enough to constitute a very undesirable residue. This was disposed of finally, after sundry tentative attempts, by coiling a piece of fine platinum wire and placing it in a porcelain vessel of water, and then leaving this fine platinum coil in constant communication with the opposite poles. The battery is unquestionably one of the most economical that has ever been set to work, considering the amount of service it is able to perform. It is calculated that the cost of maintaining the ten-celled battery in operation at the terminal stations on either side of the Atlantic, including all wear and tear, and consumption of material, will not exceed one shilling per hour.

The voltaic current therefore passes to a silk-covered wire, in innumerable coils, enveloping a bar of soft iron immediately sheathed in gutta-percha. Several miles of this fine wire (No. 20) are twined about this iron centre; then comes another coat of gutta-percha; then another coil of wire, thicker this time, (No. 14,) and 1½ miles in length. The voltaic current, passing through the wires, and reaching the iron core, converts it into a powerful magnet, exciting a current of electricity, which is delivered to the No. 20 coil, and thence to the cable, whence it departs on its Transatlantic voyage.

The transmission current generated in these double-induction coils, on reaching the further side of the Atlantic, will of course have become somewhat faint and weak from the extent of the journey it has performed. It will not, therefore, be set in this state to print or to hard work; but it will be thrown into a sort of nursery, known as the receiving instrument, where its flagging energies will be restored. The conducting strand of the cable will be here made continuous with a coil of wire, surrounding a bar of soft iron, which will become a temporary magnet, strong in proportion to the number of turns in the coil, whenever the current passes. This temporary magnet will have its precise polarity determined by the direction in which the electric current passes along the wire. The pole which will be north when the current passes in one direction, will be south when the current runs the opposite way. The apparatus relied upon by the Company to effect this object is an improvement upon the relay magnet, which figured in Messrs. Cooke and Wheatstone's patent. The advantage of it is, that the temporary magnet has no other work to do than to make the small permanent magnet traverse upon its almost frictionless pivot. On account of this peculiarity of construction, it possesses the utmost sensibility. It may be put into vigorous action by a sixpence, and a fragment of zinc placed on the moist tongue. When two or three of these instruments are scattered about in the room where the large double induction coils are at work, they are commonly heard clicking backwards and forwards automatically, and doing a little business on their own account, although no current of any kind is thrown upon their coils. They are then merely traversing upon their pivots, obediently to the magnetic attraction of the great bars, having their magnetism successively reversed some two or three yards away, and, curiously enough, are sympathetically recording, at such times, precisely the same signals and messages that the great magnets are sending off through the transmission coils.

A description of the termini of the Atlantic Telegraph line appropriately completes this history.

The American terminus is at Trinity Bay, in Newfoundland; the British, at Valentia Bay on the west coast of Ireland. The position of these two points is indicated on the accompanying map.

The approach to Trinity Bay is exceedingly picturesque, and possesses all that wildness and grandeur of scenery which distinguish nearly the whole coast of Newfoundland. When the weather is clear, the peaks of the high headlands can be seen some thirty miles out at sea, and a nearer view shows a country of peculiarly mountainous character. The first thing that strikes the visitor is the barren and rocky nature of the land; but there are some parts which are particularly susceptible of cultivation, and where, considering the inhospitable character of the climate, farming has been successfully carried on. It must be confessed, however, that the prospects for agricultural operations are not of the most encouraging kind, and that cod-fishing is, as it must always prove, the most lucrative occupation. Between those bleak, wild mountain ranges there are some beautiful little valleys, through which run streams of the purest, sparkling water. Indeed, there is an inexhaustible supply of this common but valuable and necessary article, both in the numerous lakes, rivers, and streamlets with which the whole island abounds, and which come gushing out of every rock in the summer time, or are frozen up in icy stillness and death during the long and cheerless winter.

The entrance to Trinity Bay is about thirty miles wide, and on either side rise the bold headlands of Baccalo and Horse Chops—the latter of which is about five hundred, and the former seven hundred feet in hight. The shore of the bay is marked by indentations and smaller bays, and inlets have been worn into its rocky boundaries by the restless action of the sea, which breaks here with resistless fury. Large caves, running far into the mountain barriers, have been hollowed out by the same agency, and the deep seams that scar the front of the rocks show that time has also left his mark upon them.

The other terminus of the line, is Valentia Bay, which is perhaps the most available point on the whole southern coast of Ireland, both on account of its being the nearest to Newfoundland, and on account of its particular adaptability for a telegraph station. The county of Kerry, which is indented by Valentia Bay, and in which the shore end of the cable was landed on the evening of the 6th of August, 1857, is very similar in its natural features to that part of Newfoundland which we have just described. Huge mountains rise up on almost every side, and great masses of rock, in a thousand fantastic shapes, stand out in solitary isolation miles from the land. Two of these—of such gigantic dimensions that they almost approach the dignity of mountains—guard the entrance of Dingle Bay, like weather-beaten sentinels; while further in from the ocean is a long mountain range, the face of which is worn with deep fissures, while its base is hollowed out at irregular intervals by caves, some of which extend, according to the statements of the peasantry, several hundred feet into the very heart of the mountains. The bay has a depth in some places over a hundred fathoms, but it is so open to the sea, and the anchorage is so bad, that it is one of the worst places which a vessel could select in a storm. But Valentia Bay is more protected, and although not safe in a storm, affords much better anchorage. The land, for miles into the interior, is very rocky and barren, and affords a poor pasturage for the diminutive but hardy race of cattle for which the county Kerry is so famous. The huts of the peasantry which dot the hill sides, show too plainly the poverty of the lower classes of the people, and how miserably their labor is rewarded. The ruins of churches, which were built by pious Catholics as long ago as the fourth or fifth century, are strikingly in accordance with the impoverished appearance of the land and of the people. The Island of Valentia suffered fearfully during the famine in Ireland, and hundreds died of starvation on the road side or in the miserable dwellings, some of which still remain, and in which their bodies were found many weeks after their death, unburied, and in a horrible state of emaciation. Within the last few years, it is said, the condition of the people has considerably improved.

About three miles from the head of Valentia Bay is the post town of Cahirciveen, and at the same distance, but in another direction, lies Knightstown, a small village of some five hundred inhabitants. This village is called after the Knight of Kerry, a gentleman who has been one of the strongest advocates of the Atlantic Telegraph. The small land cove in which the bay terminates was decided upon last year as the place for the landing of the cable, and has not been changed since. It is, in fact, the very best spot that could be selected. About four hundred yards from the beach, a telegraph building, somewhat similar to that at Trinity Bay, Newfoundland, has been erected, and supplied with everything necessary for the business and accommodation of the operators. The junction with the cable will be formed by a land line running to Cork, from Cork to Dublin, thence across the channel to England, and by other connexions with the great net work of telegraphs, which extends over the whole continent of Europe, and which has already embraced within it portions of Asia and Africa.

APPENDIX.

I.

ACTION OF CONGRESS IN RELATION TO THE INTERNATIONAL SUBMARINE TELEGRAPH.

IN the Senate of the United States, Dec. 23, 1856, Mr. SEWARD submitted a resolution which was unanimously adopted, requesting the President of the United States to communicate to the Senate such information as he might possess, concerning the condition and prospects of the proposed plan for connecting by submarine wires the Magnetic Telegraph wires on this continent and Europe. In response to this resolution, President PIERCE, on the 29th of the same month, transmitted to the Senate the following message and accompanying correspondence:—

To the Senate of the United States:

In compliance with a resolution of the Senate of the 23d instant, requesting the President to communicate "to the Senate, if not incompatible with the public interest, such information as he may have concerning the present condition and prospects of a proposed plan for connecting by submarine wires, the Magnetic Telegraph lines on this continent and Europe," I transmit the accompanying report from the Secretary of State.

FRANKLIN PIERCE.

Washington, *December* 29, 1856.

DEPARTMENT OF STATE,
WASHINGTON, *December* 26, 1856.

The Secretary of State, to whom was referred the resolution of the Senate of the 23d instant, requesting the President "to communicate to the Senate, if not incompatible with the public interest, such information as he may have concerning the present condition and prospects of a proposed plan for connecting, by submarine wires, the Magnetic Telegraph lines on this continent and Europe," has the honor to lay before the President a copy of a letter of the 15th instant, which he has also referred to this department, addressed to him by the President and Directors of the New York, Newfoundland, and London Telegraph Company.

W. L. MARCY.

To the President of the United States.

OFFICE OF THE NEW YORK, NEWFOUNDLAND, AND LONDON
TELEGRAPH COMPANY,
NEW YORK, *December* 15, 1856.

SIR—The undersigned, Directors of the New York, Newfoundland, and London Telegraph Company, have the honor to inform you that contracts have been made for the manufacture of the submarine telegraphic Cable, to connect the continents of Europe and America; and that it is expected to have the line between New York and London open for business by the 4th of July, 1857. A communication to this effect having been laid before the Lords Commissioners of her Britannic Majesty's Treasury, elicited a reply, of which we have now the honor to submit to you an official copy, just received by the United States' mail steamship *Atlantic*, from Cyrus W. Field, Esq., Vice-President of this Company. As the work has been prosecuted thus far with American capital, aided by the efforts of your Administration to ascertain the feasibility of the enterprise, it is the earnest desire of the directors to secure to the Government of the United States equal privileges with those stipulated for by the British Government. To this desire the Lords Commissioners of the Treasury have acceded in the most liberal spirit, by providing "That the British Government shall have a priority in the conveyance of their messages over all others, subject to the exception only of the Government of the United States, in the event of their entering into an arrangement with the Telegraph Company similar in principle to that of the British Government, in which case the messages of the two Governments shall have priority in the order in which they arrive at the stations." In view of the great international interests of this Government, and the constant recurrence of grave questions, in the solution of which time will be an essential element, we cannot doubt that the reservation made in favor of the United States will be deemed of great moment. We therefore hasten to communicate the facts to you, and to request, in view of the fact that the present Congress will soon terminate its existence, and that the Cable will be laid, if no accident prevents, before the new Congress commences its session, that you will take such action in the premises as you may deem the interests of this Government to require.

The Company will enter into a contract with the Government of the United States on the same terms and conditions as it has made with the British Government; such a contract will, we suppose, fall within the provisions of the Constitution in regard to postal arrangements, of which this is only a new and improved form.

We have the honor, also, to call your attention to the second proviso in the letter of the Lords Commissioners, to the following effect:

"Her Majesty's Government engages to furnish the aid of ships to make what soundings may still be considered needful, or to verify those already taken, and favorably to consider any request that may be made to furnish aid by their vessels in laying down the cable."

We are informed that no private steamships now built are adapted to laying a cable of such dimensions as is proposed to be used, but that the war-steamers recently finished by our Government are arranged to the very best advantage for this purpose.

To avoid failure in laying the Cable, it is desirable to use every precaution, and we therefore have the honor to request that you will make such recommendations to Congress as will secure authority to detail a steamship for this purpose, so that the glory of accomplishing what has been justly styled "the crowning enterprise of the age" may be divided between the greatest and freest Governments on the face of the globe.

With great respect, we have the honor to be, Sir, your most obedient servants,

Peter Cooper, President.
M. O. Roberts,
Moses Taylor,
Wilson G. Hunt, Directors.

The President of the United States.

TREASURY CHAMBERS, *November* 20, 1855.

SIR,—Having laid before the Lords Commissioners of her Majesty's Treasury your letter of the 13th ultimo, addressed to the Earl of Clarendon, requesting, on behalf of the New York, Newfoundland, and London Telegraph Company, certain privileges and protection in regard to the line of telegraph which it is proposed to establish between Newfoundland and Ireland, I am directed by their lordships to acquaint you that they are prepared to enter into a contract with the said Telegraph Company, based upon the following conditions, viz:

1. It is understood that the capital required to lay down the line will be (£350,000) three hundred and fifty thousand pounds.

2. Her Majesty's Government engage to furnish the aid of ships to take what soundings may still be considered needful, or to verify those already taken, and favorably to consider any request that may be made to furnish aid by their vessels in laying down the Cable.

3. The British Government, from the time of the completion of the line, and so long as it shall continue in working order, undertakes to pay at the rate of (£14,000) fourteen thousand pounds a year, being at the rate of four per cent. on the assumed capital as a fixed remuneration for the work done on behalf of the Government, in the conveyance outward and homeward of their messages. This payment to continue until the net profits of the Company are equal to a dividend of six pounds per cent., when the payment shall be reduced to (£10,000) ten thousand pounds a year, for a period of twenty-five years.

It is, however, understood that if the Government messages n any year shall, at the usual tariff rate charged to the public, amount to a larger sum, such additional payment shall be made as is equivalent thereto.

4. That the British Government shall have a priority in the conveyance of their messages over all others, subject to the exception only of the Government of the United States, in the event of their entering into an arrangement with the Telegraph Company similar in principle to that of the British Government, in which case the messages of the two Governments shall have priority in the order in which they arrive at the stations.

5. That the tariff of charges shall be fixed with the consent of the Treasury, and shall not be increased without such consent being obtained, as long as this contract lasts.

I am, sir, your obedient servant,

JAMES WILSON.

CYRUS W. FIELD, Esq., 37 Jermyn Street.

On the 9th of January, 1857, Mr. SEWARD obtained leave to introduce a bill (S. No. 493) to expedite telegraphic communication for the use of the Government in foreign intercourse; which was read twice, and referred to the Committee on the Post Office and Post Roads; and on the 13th of January, it was reported back by Mr. COLLAMER without amendment. On the 21st of January, the Senate proceeded to its consideration.

In the course of the debate, in which Senators SEWARD, HALL, RUSK, DOUGLAS, COLLAMER, TOUCEY, and others participated, Mr. SEWARD used the following remarkable language:

Mr. SEWARD: There was an American citizen who, in the year 1770, or thereabout, indicated to this country, to Great Britain, and to the world, the use of the lightning for the purposes of communication of intelligence, and that was Dr. Franklin. I am sure that there is not only no member of the Senate, but no American citizen, however humble, who would be willing to have struck out from the achievements of American invention this great discovery of the lightning as an agent for the uses of human society.

The suggestion made by that distinguished and illustrious American was followed up some fifty years afterwards by another suggestion and another indication from another American, and that was Mr. Samuel F. B. Morse, who indicated to the American Government the means by which the lightning could be made to write, and by which the telegraph wires could be made to supply the place of wind and steam for carrying intelligence.

We have followed out these suggestions of these eminent Americans hitherto, and I am sure at a very small cost. The Government of the United States appropriated $40,000 to test the practicability of Morse's suggestion; the $40,000 thus expended established its practicability and its use. Now, there is no person on the face of the globe who can measure the price at which, if a reasonable man, he would be willing to strike from the world the use of the magnetic telegraph as a means of communication between different portions of the same country. This great invention is now to be brought into its further wider and broader use—the use by the general society of nations, international use, the use of the society of mankind. Its benefits are large—just in proportion to the extent and scope of its operation. They are not merely benefits to the Government, but they are benefits to the citizens and subjects of all nations and of all States. I think there is not living in the State of South Carolina, or Tennessee, or Kentucky, or Virginia, a man who would be willing to have the use of the telegraph dispensed with or overthrown in reducing the cost of exchange of his particular products to the markets of the United States. I think so because of the celerity with which communication of the state of demand and supply in a distant market affects the value of the article in the hands of the producer, and reduces by so much the cost of the agencies employed in its sale. Precisely the same thing which thus happens at home must necessarily happen when you apply it to more remote markets in other parts of the world.

I might enlarge further on this subject, but I forbear to do so, because I know that at some future time I shall come across the record of what I have said to-day. I know that then what I have said to-day by way of anticipation, will fall so far short of the reality of the benefits which individuals, states, and nations will have derived from this great enterprise, that I shall not reflect upon it without disappointment and mortification.

At the conclusion of the debate the bill was passed, as follows:

A BILL to Expedite Telegraphic Communication for the Uses of the Government and its Foreign Intercourse.

Be it enacted by the Senate and House of Representatives of the United States of America in Congress assembled, That the Secretary of State, in the discretion and under the direction of the President of the United States, may contract with any competent person, persons, or association, for the aid of the United States in laying down a Submarine Cable, to connect existing Telegraphs between the coast of Newfoundland and the coast of Ireland, and for the use of such submarine communication, when established, by the Government of the United States, on such terms and conditions as shall seem to the President just and reasonable, not exceeding $70,000 per annum, until the net profits of such person, or persons, or association, shall be equal to a dividend of six per cent. per annum, and then not exceeding $50,000 per annum for twenty-five years: *Provided*, That the Government of Great Britain shall, before or at the same time, enter into a like contract for those purposes with the same person, persons, or association, and upon terms of exact equality with those stipulated by the United States: *And Provided*, That the tariff of prices for the use of such submarine communication by the public shall be fixed by the Secretary of the Treasury of the United States and the Government of Great Britain, or its authorized agents: *Provided further*, That the United States and the citizens thereof shall enjoy the use of the said Submarine Telegraph communication for a period of fifty years, on the same terms and conditions which shall be stipulated in favor of the Government of Great Britain, and the subjects thereof, in the contract so to be entered into by such person, persons, or association, with that Government: *Provided further*, That the contract so to be made by the British Government, shall not be different from that already proposed by that Government to the New York, Newfoundland, and London Telegraph Company, except such provisions as may be necessary to secure to each Government the transmission of its own messages by its own agents.—[Approved, March 3, 1857.]

The Charter of the Company, passed by Parliament.

The Act of Incorporation of this Company obtained at the second Session of the English Parliament for the year 1857, and receiving the Royal assent July 27 of that year, is a document of twenty-one pages. It is entitled "An Act to incorporate and regulate the Atlantic Telegraph Company, and to enable the Company to establish and work Telegraphs between Great Britain, Ireland, and Newfoundland; and for other purposes." It begins by setting forth that in October, 1856, a company was established in England to connect Great Britain or Ireland with Newfoundland by a submarine electric Telegraph, thus establishing electric communication between Europe and America, having a capital of three hundred thousand pounds divided into three hundred shares of one thousand pounds each. The number of shares was afterwards increased to three hundred and fifty, all of which were issued, and the sum of six hundred pounds was paid upon each share. It also set forth that in consequence of agreements entered into with other companies and persons, and in order that the undertaking, which is one of great public and national importance, shall be speedily carried into execution, it is desirable that a new company (including the original shareholders) should be constituted with the necessary powers to carry out the undertaking; and the Atlantic Telegraph Company was therefore incorporated, with all the rights and privileges, and assuming all the liabilities entered into by the old company. The right was granted to the new company, by a two-thirds vote, to increase the capital stock to £1,000,000, the Directors having authority to create additional shares of not more than £1,000 nor less than £20 each. The right was given to borrow one-third of the capital on bond or mortgage, but one-third of the money received for calls must be applied to the repayment of such money until the whole shall be discharged.

Ten shareholders holding stock to the amount of £15,000, have authority to require the Directors to call an extraordinary meeting of the Company, the time of the annual meeting being fixed for the month of February, in London; twenty shareholders having stock to the amount of £50,000 constituting a quorum. A two-thirds vote of the Company shall authorize the Directors to subdivide the £1,000 shares into fifty shares of £20 each, a £1,000 share conferring fifty votes upon the holder. The Board of Directors is fixed at eighteen, but the number may be reduced at any general meeting, but not below eight. The qualification of a Director is the holding of £1,000 in the stock of the Company, and the Directors of the original Company shall be the first Directors of the new Company. The remuneration of the Directors is to be fixed by the stockholders. The Company has authority to elect from the *shareholders* ordinarily resident in the United States or the British Provinces of North America, not more than eight from the former and four from the latter, who shall be Honorary Directors, shall have the right to be present, take part in and vote at the meeting of Directors, but are not to be counted in determining whether there is a quorum, and shall receive no remuneration for their services.

The British Government reserves the right to appoint an ex-officio Director of the Company, for the purpose of securing the due fulfilment on the part of the Company of all contracts for the transmission of signals and messages for her Majesty or on her Majesty's service. This ex-officio Director is not to go out of office with the other Directors, but he is removable at the pleasure of the Government. He is to be a shareholder in the Company or not, as the Government may think fit; he is to be present at all meetings of the Directors and of the Company; and has power to examine all books and documents of the Company; but has not the right to vote, and does not receive any remuneration from the Company.

He has the power, in case he is of opinion that any act or course of the Company is prejudicial to the performance of the contracts with Government, or the regular, speedy, and impartial transmission of messages for the public, or otherwise disadvantageous to Government or the public, to veto the taking of such course or the doing of such act, when the matter shall be referred to the Lords Commissioners of the Treasury, whose opinion shall be final, unless the Board of Directors see fit to appeal to two Judges of the Superior Court at Westminster, whose decision shall be final and conclusive on all parties. The election of Directors of the Company, ordinary or honorary, is subject to the approval of the Lords Commissioners of the Treasury.

The undertaking of the Company was set forth to be the laying down of one or more submarine Telegraph Cables between Ireland and Newfoundland, or on the Continent of America, and the working of such lines. They are authorized and empowered to make arrangements with the New York, Newfoundland, and London and other Companies, for the transmission of messages and for the charges on the same, as may be necessary for the carrying out of the object of the Company. They are empowered to fix and receive reasonable charges for the transmission of messages, and may demand pre-payment of the same. With the exception of the priority of government messages, all others for the public are to be received and sent without favor or preference, according to the order of time in which they shall have been received by the Company. The following section in regard to the right of government priority, we quote entire:—

"LVI. All messages and signals sent or forwarded for transmission and delivery for her Majesty or on Her Majesty's service shall have priority over all other messages whatsoever, and it shall be imperative on the Company, their officers and servants, to transmit and deliver such messages and signals accordingly, and to suspend the transmission of all or any other messages until the said messages and signals shall first have been transmitted; Provided, always, that the Company may, in consideration of a guarantee or subsidy granted or secured by the Government of the United States, equal in rate or amount to that granted by or on behalf of her Majesty's Government, grant and extend to the Government of the United States the like priority for intelligence, on and for their service over all other messages and signals whatsoever, except those for her Majesty or on her Majesty's service, and after they shall have so done, and shall have notified their having so done to the Lords Commissioners of her Majesty's Treasury, the messages and signals on the service of the Government of the United States shall thenceforward be entitled to, and shall have, during the continuance in force of any such guaranty or subsidy, the like priority as messages and signals for her Majesty or on her Majesty's service over those of all other persons whomsoever, and thenceforward messages and signals for her Majesty or on her Majesty's service, and those on the service of the Government of the United States, shall, as between themselves, have no right of priority, but be transmitted and delivered respectively in the order of time in which they may be respectively tendered for transmission and delivery."

If the Government and the Company cannot agree upon the rate of remuneration for sending public messages, the matter is to be settled by referees.

The Company is authorized, whether or not they shall grant for the Government of the United States any such priority as is stated in the section before quoted, to make arrangement with that Government for the transmission of their messages. The following section authorizes the English Government, under certain circumstances, to take possession of the works of the Company.

"LXII. At all times from and after the period of twenty-five years from the opening of the said lines of Telegraph communication for the transmission of messages, whenever one of her Majesty's principal Secretaries of State for the time being shall be of opinion that circumstances render it expedient to vest in her Majesty's Government the control of the operations of the Company, it shall be lawful for such Secretary of State, by warrant under his hand, to cause possession to be taken of all the Telegraphs and Telegraphic apparatus of the various stations of the Company, their licenses or assigns, for the space of one week from the date of such warrant, for the purpose of preventing any communication being made or signals given, save such as shall be directed and authorized by any such Secretary of State, and also by further successive warrants to cause possession of the said Telegraphs and Telegraphic apparatus to be retained from week to week, so long as any such Secretary of State shall deem such possession expedient for the public service: Provided always, that for every week during which possession shall be so retained, the Company, their licensees or assigns, shall receive from and be paid by the Lords Commissioners of her Majesty's Treasury, the same amount of profits as the Company would have made in case they had continued the working of the said Telegraphs, such profits to be computed upon an average of the weekly profits of the Company for the three months immediately preceding the issuing of the first of the said warrants."

Any negligence or delay in the transmission of a message, makes the officer liable to a fine not exceeding £20. Any damage to the works of the Company is recoverable by a suit at law, and any person committing such injury is liable also to punishment as for the commission of a misdemeanor.

II.

LIEUT. M. F. MAURY ON THE FEASIBILITY OF OCEANIC TELEGRAPHS.

On the 22d of February, 1854, more than four years and a half ago, Lieut. Maury addressed the following letter to the Secretary of the Navy. Its predictions have been signally verified:

National Observatory,
Washington, *February* 22, 1854.

Sir,—The United States brig Dolphin, Lieutenant Commanding O. H. Berryman, was employed last summer upon especial service connected with the researches that are carried on at this office concerning the winds and currents of the sea. Her observations were confined principally to that part of the ocean which the merchantmen, as they pass to and fro upon the business of trade between Europe and the United States, use as their great thoroughfare. Lieutenant Berryman availed himself of this opportunity to carry along also a line of deep sea soundings from the shores of Newfoundland to those of Ireland. The result is highly interesting, in so far as the bottom of the sea is concerned, upon the question of a Submarine Telegraph across the Atlantic; and I therefore beg leave to make it the subject of a special report.

This line of deep sea soundings seems to be decisive of the question as to the practicability of a Submarine Telegraph between the two continents, *in so far as the bottom of the deep sea is concerned.* From Newfoundland to Ireland, the distance between the nearest points is about 1,600 miles;* and the bottom of the sea between the two places is a plateau, which seems to have been placed there especially for the purpose of holding the wires of a Submarine Telegraph, and of keeping them out of harm's way. It is neither too deep nor too shallow; yet it is so deep that the wires but once landed, will remain for ever beyond the reach of vessels' anchors, icebergs, and drifts of any kind, and so shallow that the wires may be readily lodged upon the bottom. The depth of this plateau is quite regular, gradually increasing from the shores of Newfoundland to the depth of from 1,500 to 2,000 fathoms as you approach the other side. The distance between Ireland and Cape St. Charles, or Cape St. Lewis, in Labrador, is somewhat less than the distance from any point of Ireland to the nearest point of Newfoundland. But whether it would be better to lead the wires from Newfoundland or Labrador is not now the question; nor do I pretend to consider the question as to the possibility of finding a time calm enough, the sea smooth enough, a wire long enough, a ship big enough, to lay a coil of wire 1600 miles in length; though I have no fear but that the enterprise and ingenuity of the age, whenever called on with these problems, will be ready with a satisfactory and practical solution of them.

I simply address myself at this time to the question in so far as *the bottom of the sea* is concerned, and as far as that the greatest practical difficulties will, I apprehend, be found after reaching soundings at either end of the line, and not in the deep sea. * *

A wire laid across from either of the above-named places on this side will pass to the north of the Grand Banks, and rest on that beautiful plateau to which I have alluded, and where the waters of

* From Cape Freels, Newfoundland, to Erris Head, Ireland, the distance is 1,611 miles; from Cape Charles, or Cape St. Lewis, Labrador, to ditto, the distance is 1,601 miles.

the sea appear to be as quiet and as completely at rest as it is at the bottom of a mill-pond. It is proper that the reasons should be stated for the inference that there are no perceptible currents, and no abrading agents at work at the bottom of the sea upon this Telegraphic Plateau. I derive this inference from a study of a physical fact, which I little deemed, when I sought it, had any such bearings.

Lieut. BERRYMAN brought up with BROOKE's deep-sea sounding apparatus specimens of the bottom from this plateau. I sent them to Prof. BAILEY, of West Point, for examination under his microscope. This he kindly gave; and that eminent microscopist was quite as much surprised to find, as I was to learn, that all those specimens of deep-sea soundings are filled with microscopic shells; to use his own words, "*not a particle of sand or gravel exists in them.*" These little shells, therefore, suggest the fact that there are no currents at the bottom of the sea whence they came—that BROOKE's lead found them where they were deposited in their burial-place after having lived and died on the surface, and by gradually sinking were lodged on the bottom. Had there been currents at the bottom, these would have swept and abraded and mingled up with these microscopic remains the debris of the bottom of the sea, such as ooze, sand, gravel, and other matter; but not a particle of sand or gravel was found among them. Hence the inference that these depths of the sea are not disturbed either by waves or currents. Consequently, a telegraphic wire once laid there, there it would remain, as completely beyond the reach of accident as it would be if buried in air-tight cases. Therefore, so far as the bottom of the deep sea between Newfoundland, or the North Cape, at the mouth of the St. Lawrence, and Ireland, is concerned, the practicability of a Submarine Telegraph across the Atlantic is proved. * *

In this view of the subject, and for the purpose of hastening the completion of such a line, I take the liberty of suggesting for your consideration the propriety of an offer from the proper source, of a prize to the Company through whose Telegraphic Wire the first message shall be passed across the Atlantic.

I have the honor to be, respectfully, &c.,

M. F. Maury, Lieut. U.S. Navy

Hon. J. C. Dobbin, Secretary of the Navy, Washington, D.C.

In December, 1856, the following correspondence passed:—

House of Representatives,
Washington, *December* 30, 1856.

Sir,—The submarine communication which now excites so much attention, both in the Congress of the United States and the country, will, I perceive by the map of the survey, terminate on this side the Atlantic in the British possessions, *i.e.* in Newfoundland.

Will you do me the favor, at your earliest convenience, to answer the following questions, to wit:

Is there a point, *under our flag*, which would answer for the western terminus?

If not, what are the obstructions?

What influence would it have in a military point of view?

Very respectfully, your obedient servant,

C. C. Chaffee.

Lieut. Maury, U.S. Navy.

U.S.N. Observatory and Hydrog. Office,
Washington, *December* 31, 1856.

Sir,—I have received your note of the 30th instant, making certain inquiries in relation to the Submarine Telegraph of the Atlantic, and wishing to know what are the obstructions which prevent the western end of the wire from being brought straight across the sea to our own shores.

The difficulties are manifold, and in the present state of the telegraphic art, they may be considered insuperable.

The shortest telegraph distance between the British Islands and the United States, without touching English soil by the way, is, in round numbers, three thousand miles, and the lightning has never yet been made to bear a message through a continuous wire of such a length. Here, therefore, is an obstruction.

The distance from the Western Islands to the nearest point on our shores is about equal to the distance between Newfoundland and Ireland; and the distance between the Irish coast and the Western Islands is about fifteen hundred miles. Therefore, with a relay on the Western Islands a line from Ireland, via those Islands to our own shores, is electrically practicable.

But a wire by that route would have to cross the Atlantic at its deepest part, and then the Portuguese Government, as well as the English, would have control of the line; so that, in a military, commercial, or political point of view, nothing would be gained by underrunning the Atlantic with the telegraphic wires by that route. Moreover, that route would lead the wires across a volcanic region. These constitute obstructions that, in the present state of our knowledge, are fatal to such a route.

The only practicable route for a Submarine Telegraph between the United States and England appears to be along the "plateau" of the Atlantic, whereon it is proposed to lay the wire that is now in process of construction.

But suppose a line were to be constructed by American enterprise from the British shores, all the way to one of our sea-port towns: *cui bono?* In time of peace the line along the "plateau" would, by reason of its great advantages, take all the business; and in war the British authorities need but cut the American cord, or take charge of its office at the other end, to render the whole line inoperative or perfectly useless to us.

It cannot but be regarded by every wise and good man as a fortunate circumstance, that this great enterprise of the sub-Atlantic Telegraph is the joint work of England and America. This circumstance ought of itself to serve as a guarantee to the world that, in case of war—should war unhappily ever be waged between these two nations—that cord is never to be broken, or to be used otherwise than freely and fairly alike by the two nations, their citizens and subjects.

We have just seen the great nations of Europe emerging from the horrors of a fierce and bloody war; and yet, to their honor and the glory of the age be it said, that that strife, vengeful though it was, was not savage enough to break a single line of telegraphic wire. The lightning ran to and fro with messages between St. Petersburgh and the capitals of France and England, as it now does. And in case of war with this country, after that electric cord is stretched by the joint means and enterprise of the two people upon the quiet bottom of the deep sea, neither of the two Governments would dare take that cord, and, in the face of the Christian states and people of the age, convert it into a military engine to be turned against its joint-owners and partners.

Our fellow-citizens who contrived, planned, and brought forward this noble work, are too sagacious and patriotic not to have perceived that, lying as it does wholly within the control of a foreign power, that power, were it a nation of Goths and Vandals, might turn the path they were about to make for the lightning along the bed of the ocean against their own country in war; but they knew the people on the other side, and trusted to higher and nobler sentiments. The British Government interfere with the free use of that Cable even in war! The spirit of the age is against such an act, and no State within the pale of Christendom, much less that great English nation of noble people, would dare to do such a thing. Her people and rulers would not if they could; they could not if they would. We might as well think of tearing up now, in peace, the railways between Canada and the States, or of abrogating the steam-engine because it may be turned against us in war.

When Captain Cook was on his voyage of discovery, France and England were at war. The king of France was requested not to let his armed cruisers destroy the records of that expedition in case any of them should fall in with it. You recollect the noble reply: "I war not against science;" and forthwith every French man-of-war had orders to treat Cook as a friend, should they fall in with him; and assist, not interrupt, him in the object of his cruise. To this day the memory of that king is held in more esteem for that act and sentiment than for any other act of his reign.

A little more than three years ago, at the maritime conference of Brussels, where the principal nations of the world assembled in the persons of their representatives, to devise a uniform plan of physical research at sea, and to report the best form for the abstract log to be used on board ship for marking the observations upon its winds and currents, those functionaries alluded to this sentiment of the French monarch, and appealed each to his own Government to order that, in case of war, this abstract log should also be regarded as a sacred thing. It is made so. The armed cruisers of the various nations that are co-operating in this system of research are required to touch that record with none but friendly hands.

This Submarine Telegraphic line is an achievement which this very system of research has had something to do in bringing about; and is it likely that it will or can be monopolized by any power for war purposes? Fairly and clearly it may be considered as the joint property of those who are operating as co-workers and joint co-laborers in that beautiful system of physical research by which a way for the lightning has been discovered under the sea and across the ocean.

This system of research, it has been proclaimed over and over again, was not undertaken for the exclusive advantage of any one people or nation, but for the benefit of commerce, the advancement of science, and for the benefit and improvement of the whole human family; and with this understanding the nations of Europe entered into it.

Being joint owners and equal participators in such a great enterprise as this, we may, with propriety, under these circumstances, demand a fair participation in all its advantages.

But suppose we should stand aloof, and that the enterprise now on foot should be abandoned by our citizens and government, and then suppose war to come; in less than six months after its declaration, the British government could, on its own account, have a wire stretched along this telegraphic plateau between Newfoundland and Ireland.

You do not desire me in your note to consider the Christianizing, political, social, and peace-preserving influences which this fascicle of copper threads, when once stretched upon the bed of the ocean, is to have, and therefore I do not offer any of the views which present themselves from such a stand-point. This much, however, I may say: Submarine Telegraphy is in its infancy, but it is in the act of making the stride of a full-grown giant; and no problem can to my mind be more satisfactorily demonstrated than is the practicability of readily, and almost without risk, laying the wire from land to land upon this telegraphic plateau of the Atlantic.

Respectfully, &c., M. F. Maury.

Hon. C. C. Chaffee, House of Representatives, Washington.

III.

THE BASIN OF THE ATLANTIC, AND THE TELEGRAPHIC PLATEAU.*

THERE is at the bottom of the Atlantic, between Cape Race in Newfoundland and Cape Clear in Ireland, a remarkable steppe, which is already known as the telegraphic plateau. A Company is now engaged with the project of a submarine telegraph across the Atlantic. It is proposed to carry the wires along this plateau from the eastern shores of Newfoundland to the western shores of Ireland. The great circle distance between these two shore-lines is one thousand six hundred miles, and the sea along the route is probably nowhere more than ten thousand feet deep. This Company, it is understood, consists of men of enterprise and wealth, who, should the inquiries that they are now making prove satisfactory, are prepared to undertake the establishment forthwith of a Submarine Telegraph across the Atlantic.

It was upon this plateau that BROOKE's sounding apparatus brought up its first trophies from the bottom of the sea. These specimens Lieutenant BERRYMAN and his officers judged to be clay; but they took the precaution to label them, carefully to preserve them, and, on their return to the United States, to send them to the proper bureau. They were divided: a part was sent for examination to Professor EHRENBERG, of Berlin, and a part to Professor BAILEY, of West Point—eminent microscopists both. I have not heard

* From the "Physical Geography of the Sea," by Lieut. M. F. MAURY. New York: Harper & Brothers, 1855.

from the former, but the latter, in November, 1853, thus responded:

"I am greatly obliged to you for the deep soundings you sent me last week, and I have looked at them with great interest. They are exactly what I have wanted to get hold of. The bottom of the ocean at the depth of more than two miles I hardly hoped ever to have a chance of examining; yet thanks to BROOKE'S contrivance, we have it clean and free from grease, so that it can at once be put under the microscope. I was greatly delighted to find that all these deep soundings were filled with microscopic shells; not a particle of sand or gravel exists in them. They are chiefly made up of perfect little calcareous shells (*Foraminiferæ*), and contain, also, a small number of silicious shells (*Diatomaceæ*)."

These little mites of shells seem to form but a slender clue indeed by which the chambers of the deep are to be threaded, and mysteries of the ocean revealed; yet the results are suggestive; in right hands and to right minds, they are guides to both light and knowledge. The first noticeable thing the microscope gives of these specimens is, that all of them are of the animal, not one of the mineral kingdom. The ocean teems with life, we know. Of the four elements of the old philosophers—fire, earth, air, and water—perhaps the sea most of all abounds with living creatures. The space occupied on the surface of our planet by the different families of animals and their remains are inversely as the size of the individual. The smaller the animal, the greater the space occupied by his remains. Though not invariably the case, yet this rule, to a certain extent, is true, and will, therefore, answer our present purposes, which are simply those of illustration. Take the elephant and his remains, or a microscopic animal and his, and compare them. The contrast, as to space occupied, is as striking as that of the coral reef or island with the dimensions of the whale. The grave-yard that would hold the corallines is larger than the grave-yard that would hold the elephants.

We notice another practical bearing in this group of physical facts that BROOKE's apparatus fished up from the bottom of the deep sea. BAILEY, with his microscope, could not detect a single particle of sand or gravel among these little mites of shells. They were from the great telegraphic plateau, and the inference is that there, if anywhere, the waters of the sea are at rest. There was not motion enough there to abrade these very delicate organisms, nor current enough to sweep them about and mix up with them a grain of the finest sand, nor the smallest particle of gravel torn from the loose beds of debris that here and there strew the bottom of the sea. This plateau is not too deep for the wire to sink down and rest upon, yet it is not so shallow that currents, or icebergs, or any abrading force can derange the wire, after it is once lodged.

As Professor BAILEY remarks, the animalcula, whose remains BROOKE's lead has brought up from the bottom of the deep sea, probably did not live or die there. They would have had no light there, and had they lived there, their frail little textures would have been subjected in their growth to a pressure upon them of a column of water twelve thousand feet high, equal to the weight of four hundred atmospheres. They probably lived and died near the surface, where they could feel the genial influences of both light and heat, and were buried in the lichen caves below after death.

BROOKE's lead and the microscope, therefore, it would seem, are about to teach us to regard the ocean in a new light. Its bosom, which so teems with animal life; its face, upon which time writes no wrinkles—makes no impression—are, it would now seem, as obedient to the great law of change as is any department whatever, either of the animal or the vegetable kingdom. It is now suggested that, henceforward we should view the surface of the sea as a nursery teeming with nascent organisms, its depths as the cemetery for families of living creatures that outnumber the sands on the sea-shore for multitude.

Where there is a nursery, hard by there will be found also a grave-yard—such is the condition of the animal world. But it never occurred to us before to consider the surface of the sea as one wide nursery, its every ripple a cradle, and its bottom one vast burial-place.

On those parts of the solid portions of the earth's crust which are at the bottom of the atmosphere, various agents are at work, levelling both upward and downward. Heat and cold, rain and sunshine, the winds and the streams, all assisted by the forces of gravitation, are unceasingly wasting away the high places on the land, and as perpetually filling up the low.

But in contemplating the levelling agencies that are at work upon the solid portions of the crust of our planet which are at the bottom of the sea, one is led at first thought almost to the conclusion that these levelling agents are powerless there.

In the deep sea there are no abrading processes at work; neither frosts nor rains are felt there, and the force of gravitation is so paralysed down there that it cannot use half its power, as on the dry land, in tearing the overhanging rock from the precipice and casting it down into the valley below.

IV.

EARLY PREDICTIONS OF PROFESSOR MORSE.

NEW YORK, *August* 10*th*, 1843.

SIR: I take this opportunity of communicating to the honorable Secretary the result of the experiments made on the 8th inst., with the prepared wire in one continuous line of 160 miles. Professors Renwick, Draper, Ellet, and Schaeffer, with my assistants Professors Fisher and Gale, were present by invitation; Professors Silliman, Henry, Torrey, and Dr. Chilton were also invited, but were prevented by official duties from attending.

In the letter to the honorable Secretary dated March 10th, 1843, in which I propose my general plan, I have this remark, speaking of the wire after its insulating preparation should be completed: "Many interesting *experiments* bearing upon the general result can then be tried before the wire is enclosed."

The experiments alluded to were tried on Tuesday, and with perfect success. I had prepared a galvanic battery of 300 pairs in order to have ample power at command, but, to my great gratification, I found that 100 pairs were sufficient to produce all the effects I desired through the whole distance of 160 miles.

It may be well to observe that the 160 miles of wire are to be divided into four lengths of 40 miles each, forming a fourfold cord from Washington to Baltimore. Two wires form a circuit; the electricity, therefore, in producing its effects at Washington from Baltimore, passes from Baltimore to Washington and back again to Baltimore, of course travelling 80 miles to produce its result. One hundred and sixty miles, therefore, gives me an actual distance of 80 miles, double the distance from Washington to Baltimore. The result, then, of my experiments on Tuesday is, that a battery of only 100 pairs at Washington will operate a telegraph on my plan 80 miles distant with certainty, and without requiring any intermediate station!

Some careful experiments on the decomposing power at various distances were made, from which the law of propulsion has been deduced, verifying the results of Ohm and those which I made in the summer of 1842, and alluded to in my letter to the Hon. C. G. Ferris, and published in the House report No. 17 of the last Congress.

The practical inference from this law is that a telegraphic communication on the electro-magnetic plan may, with certainty, be established across the Atlantic ocean! Startling as this may now seem, I am confident the time will come when this project will be realized.

The wire is now in its last process of preparation for enclosing in the lead tube, which will be commenced on Tuesday, the 15th inst.

I have the honor to be, sir, with sincere respect, your most obedient servant,

SAMUEL F. B. MORSE,

Superintendent of Electro-Magnetic Telegraph.

To the Hon. John C. Spencer,

Secretary of the Treasury of the U. States.

V.

USE OF THE TELEGRAPH IN CONNEXION WITH LONGITUDE OPERATIONS.

At the Second Annual Meeting of the American Association for the Advancement of Science, held at Cambridge, Mass., in August, 1849, a valuable paper, embodying the results of Telegraphic observations of Longitude, was read by Mr. Sears C. Walker, a capable astronomer, since deceased. The investigations upon this subject were undertaken by Mr. Walker, under the direction of the Superintendent of the Coast Survey. The question of determining longitudes by the use of telegraphic wires, is now invested with fresh interest. The material portions of the report of Mr. Walker may, therefore, be reproduced in connection with this history. Mr. Walker said:

"The first mention of the electro-magnetic telegraph, in connection with longitude operations, as far as I know, was made, in 1837, by M. Arago to Dr. Morse.

"The first practical application of the method was by Capt. Wilkes, in 1844, between Washington and Baltimore. Two chronometers, previously rated by astronomical observations in the vicinity, were brought to the two telegraph offices, and were compared together through the medium of the ear, without coincidence of beats. This process is accurate enough for geographical or nautical purposes: but its precision stops short of the mark where the requirements of geodesy begin. In fact, two clocks beating the same kind of time, when placed side by side, cannot be compared together, by the human ear, with sufficient precision for geodetical purposes. The subsequent experience of the Coast Survey has shown, that where several astronomers make independent comparisons of clocks, in this manner, two seconds of an arc, or twelve hundredths of a second of time, is an average discrepancy between their results.

"The subject of telegraph operations for longitude had engaged the attention of the Superintendent of the Coast Survey previous to the experiment of Capt. Wilkes; but the orders received by me for this purpose bear date November 24, 1845. In 1846, the very first season in which two astronomical stations of the Survey were brought in connection by the Morse telegraph lines, the. work of connecting them together in longitude was commenced in earnest by the superintendent of the Coast Survey. The coöperation of the National Observatory, as one of the stations, was freely tendered by its Superintendent, Lieut. Maury, U. S. N., and accepted by Prof. Bache.

"Another station was established at Philadelphia, under the superintendence of Prof. Kendall, and still another at Jersey City under Prof. Loomis.

"Owing to the imperfect insulation of the lines, the connection of Jersey City with Washington failed that year; but the Washington and Philadelphia stations were connected together astronomically on the 10th and 22d of October. The method of comparison by coincidence of beats of solar and sidereal timekeepers, was not introduced this year; but the equivalent one was employed, viz., the exchange of star-signals. These are the dates of instants of the passage of a star over the wires of the eye-piece of the transit instrument, signalized by tapping on the telegraph key at one station, and recording it on the Morse register at both.

"In 1846, we connected together in longitude the Washington and Philadelphia stations. In 1847, the programme left unfinished in 1846, by the imperfection of the lines, was resumed and completed, and Washington, Philadelphia, and Jersey City were connected together. On the 27th of July, 1847, the method of coincidence of beats, used so successfully by R. T. Paine, Esq., in the chronometric operations for longitude in Massachussets, and by Struve and Airy in their chronometric enterprises, was applied to the telegraphic comparisons of the Philadelphia and Jersey City clocks. This method of coincidences was used in combination with exchanges of star-signals in the telegraphic operations of the Coast Survey in 1848, when the Cambridge Observatory, under Prof. Bond, and the Stuyvesant Station in New York, were connected together by the Coast Survey.

"In October, 1848, Cincinnati was connected with Philadelphia. The labors of the year 1848 comprise some 1,800 observed transits of stars, 800 comparisons of chronometers by coincidences of beats taken at the stations, 5,000 transits over wires, for determining the personal equations of the officers of the Survey, many thousand exchanges of personal clock signals, and 600 star-transit signals.

"Of the different kinds of registers I prefer the sheet of Mr. Saxton. One sheet filled on both sides, or two pages, will contain an ordinary night's work. A year's work will make a book of some three hundred pages, on the margin of which may be entered the ordinary remarks for an observing-book, relative to the state of the level and meteorological instruments, names of stars observed, and instrumental deviations. If folded up, or bound and put away for a century, the reduction of the work will then be as easy as at first. In fact, we may, with the metallic cylinder, electrotype the plate; or, using copper, we may print from it without.

"When we reflect that the probable error of one transit over one wire is only the sixteenth of a second, and that with five wires it is only a thirty-sixth part, or three hundredth of a second, it is manifest that one tally, or five wires, is ample for all ordinary work. In fact, one wire is sufficient for most of the purposes of astronomy. I have been led, on consideration of all the facts known from the experience of the Coast Survey, to make the following remark relative to the precision of our work, after proper adjustment of the transit instrument, or measurement of its deviations from a normal state :—*The printed transit of a fundamental star over any one wire of Wurdeman's diaphragm, and that of a star, planet, or comet, whose place is sought, over another wire,—both reduced to the centre, on the supposition of uniformity of interval,—give the place of the object sought with a precision not much below that on which rest the present elements of all the bodies in the solar system.*"

VI.

VELOCITY OF THE GALVANIC CURRENT.

In the year 1850, the American Association received a paper embodying the results of experiments on the velocity of the Galvanic Current in Telegraph wires. These experiments were made under the direction of Prof. A. D. Bache, Superintendent of the Coast Survey; and a very excellent summary of them was prepared and laid before the American Association, by Dr. B. A. Gould, Jr., from whose report we copy:—

The ingenious experiment* of Prof. Wheatstone, in 1834, tended to confirm the general opinion previously existing, that the velocity with which the electricity was transmitted by a metallic conductor was so enormous—so immense, indeed, compared with all other velocities known to us, excepting that of light—as to warrant the assumption of our incapacity to determine it. On this account, Wheatstone's elegant experiment obtained for its author the more distinction, and for his results the greater confidence. One of these results, as announced by him, was, that the velocity of electricity through the copper wires used, was indeed appreciable,—but exceeded that of light through the planetary space,† that it could not be less than 288,000 miles in a second, while light traverses about 186,000 during the same time.

The telegraphic observations, instituted under the immediate direction of Mr. Walker, by the U. S. Coast Survey, for determining the differences of longitude between remote stations in the

* Phil. Trans., 1834, p. 583.

† Phil. Trans., 1834, p. 591.

United States, led to a very unexpected result,—viz.: that to obtain the greatest harmony among the several observations, a small correction must be introduced, depending on the relative distances between the telegraphic stations. No explanation of this phenomenon offered itself, excepting the hypothesis suggested* by Walker, and communicated by Prof. Bache to the Am. Phil. Society in March, 1849, that the time elapsing during the passage of the signals between remote stations was much more considerable, and the velocity, consequently, less than had been before imagined.

Since Walker's results were first published, the subject has engaged the attention of numerous astronomers and physicists in Europe and America, among whom Mitchel, Fizeau, and Steinheil are conspicuous. The subject belongs in itself far more properly to the domain of physics than to astronomy, but its special bearing upon the problem of longitude, and the manner in which it has forced itself upon the consideration of astronomers, have made it incumbent upon them to enter into a full discussion of the subject.

While in Washington in the month of February last, I accepted with pleasure an invitation from Mr. Walker to take part in an experiment on a very large scale, for which he had been long engaged in making preparation in behalf of the Coast Survey, and from which he anticipated results so ample as to put an end to the controversy. The Seaton Station of the Survey in Washington, north of the Capitol, and the city of St. Louis, were connected on the 4th February, in one colossal galvanic circuit, and but for the damage occasioned by a storm on the same day, the circuit would have extended even to Dubuque, in the Territory of Iowa, a distance of some 1500 miles.

On the night of Oct. 31, 1849, a series of experiments was made

* Proc. Am. Phil. Soc., v. p. 76. Astr. Nachr., xxix. 54.

for the express purpose of determining the time needed for the transmission of signals. The results are published in No. 7 of the *Astr. Journ.*, with a detailed account of the methods which he used, and an analytical investigation of the effects of those circumstances which could interfere with the accuracy of his results. The measurements of all the registers gave him for the velocity on that night 16,000 miles a second,—differing less than 1900 miles from his previous result, and tending in general to confirm it. The final result at which he arrived was the general theorem,—that a signal given by breaking or closing the galvanic circuit at any point, was observed at other points on the circuit after intervals proportionate to their distance from the place where the signal was made,—and corresponding to a velocity of from 16,000 to 19,000 miles.

Prof. Mitchel, of the Cincinnati Observatory, dissents from the view taken by Mr. Walker, and attributes the results obtained by him to the effect of various sources of error and uncertainty in the methods which Walker has used. He devised a special and very curious apparatus for investigating the question,—and with the ingenuity and mechanical skill for which he is so eminent, constructed it at the Cincinnati Observatory, and made a large series of interesting experiments on the Telegraph line between Cincinnati and Pittsburgh. Prof. M.'s view of the matter is, that after a signal is given by closing or breaking the galvanic circuit, an appreciable time elapses before the signal is communicated to any other station, and that it is then received by all simultaneously. He considers this in connection with the details of his experiment to indicate that two fluids circulate in opposite directions between the poles of a battery, but that neither makes its influence perceptible until complete circulation of each has taken place from pole to pole. The velocity of this circulation Prof. Mitchel infers to be about 30,000 miles a second.

Were the arrangement of the Telegraph such now as it was at first, one source of uncertainty would perhaps have been avoided in the experiments; but the opportunity of solving this latter problem would have been lost. Until telegraphers availed themselves of the discovery of Steinheil, that no control over the circuit was lost when one half of it was formed by the earth, each Telegraph line was double—consisting of one wire to the terminus and another back. But in all the lines in use in this country, the earth forms one half of the circuit. Are we to consider, when the two distant extremities of a line of wire communicate with the earth at a distance of many hundred miles from one another, that there is a special line of tension through the earth from one extremity to the other? and that a signal is communicated from terminus to terminus through the ground, in the same manner as it is through a wire? or may we consider the earth as a huge receptacle, to speak metaphorically, capable of receiving or imparting any amount of electricity at any time? The former opinion is held by my friend Mr. Walker.

But does it not seem improbable that the slight activity of a galvanic battery, traversing a circuit of 1000 miles of wire, should be sufficient to establish a *special* line of electric tension extending through the earth in a cord or parallel with the surface for 750 miles? For my own part, when I remember not only the grand phenomena of terrestrial magnetism, but the immense galvanic force which must be exerted by the mutual influence of the huge masses of metal in the bowels of the earth,—when I consider the mighty electrical activity developed in the great processes of nature, —I will confess that I cannot bring myself to believe that *one special continuous line* of electric tension in the ground between two remote stations can be established athwart all these colossal forces by the action of a puny Telegraph battery.

VII.

TABLE OF SUBMARINE CABLES.

Route.	Date.	Miles.
Dover and Calais,	1850	24
Dover and Ostend,	1852	76
Holyhead and Howth,	1852	65
England and Holland,	1853	115
Portpatrick and Donaghadee (two Cables),	1853	26
Italy and Corsica,	1854	65
Corsica and Sardinia,	1854	10
Denmark—Great Belt,	1854	15
Denmark—Little Belt,	1854	5
Denmark—Sound,	1855	12
Scotland—Frith of Forth,	1855	4
Black Sea,	1855	400
Soland, Isle of Wight,	1855	3
Straits of Messina,	1856	5
Gulf of St. Lawrence,	1856	74
Straits of Northumberland,	1856	10½
Bosphorus,	1856	1
Gut of Canso, Nova Scotia,	1856	2
St. Petersburg to Cronstadt,	1856	10
Atlantic Cable—Valentia Bay to Trinity Bay,	1858	1950
		2,862½

VIII.

THE MORSE TELEGRAPHIC ALPHABET.

Letters.				Figures and Punctuation.	
A	-—	O	- -	1	-——-
B	—---	P	-----	2	--—--
C	-- -	Q	--—-	3	---—-
D	—--	R	- --	4	----—
E	-	S	---	5	———
F	-—-	T	—	6	------
G	——-	U	--—	7	——--
H	----	V	---—	8	—----
I	--	W	-——	9	—--—
J	—-—-	X	-—--	0	———
K	—-—	Y	-- --	(.)	--——--
L	——	Z	--- -	(?)	—--—-
M	——	&	- ---	(!)	————
N	—-	&c	- --- ---		

IX.

RECEPTION OF THE TIDINGS OF SUCCESS IN THE UNITED STATES.

The tidings of the mechanical success of the enterprise, first received in New York on the 5th of August, gave the first impulse to a universal jubilee. That the Cable had been safely laid, was a fact which justified a warm expression of popular feeling. The successful laying of an Oceanic Telegraph was justly regarded by the American people as an achievement which carried rich compensation for previous trials and dangers. For the space of a week, therefore, while the question of overcoming the scientific difficulties of transmission of the electric current yet remained unsettled, the country gave itself up to a general jubilation.

The announcement of the landing of the Cable at Trinity Bay reached the quiet town of Andover, Massachusetts, while the Alumni of the Theological Seminary at that place were celebrating their semi-centennial anniversary by a dinner. One thousand persons were present, all of whom rose to their feet, and gave vent to their excited feelings by continued and enthusiastic cheers. When quiet was restored, the Alumni sang the Doxology to the tune of "Old Hundred;" brief addresses, referring to the great event as a new link in the influences of Christianity, were delivered by the Rev. Wm. Adams, D.D., of New York, the Rev. Dr. Hawes of Hartford, and others; a prayer was offered, and the dinner was then resumed.

The news was received at Washington with unbounded enthusiasm. Labor was entirely suspended in the Government Departments, and the tidings spread over the city with extraordinary rapidity. The President was not in town, having retired, some days previously, to Bedford, Pennsylvania, at which place the announcement was made to him.

At Albany, New York, the proceedings of the Courts, of the Board of Trade, and of the Railroad Companies, were instantly arrested, and intense excitement prevailed. In that city, and at Utica, Syracuse, Rochester, Buffalo, and other cities and towns in the State of New York, public and private buildings were illuminated, and enthusiasm found vent in various demonstrations.

In the city of New York, the first news was received with some degree of caution. Hence the celebration of the event on the night of the 5th of August was not equally enthusiastic with the demonstrations elsewhere. The city reserved itself for a future occasion. On the reception of the Queen's Message, its exuberance knew no bounds.

At Fishkill, New York, an impromptu celebration took place, at which speeches were delivered by the Rev. Henry Ward Beecher, and others. In the course of his remarks, Mr Beecher said:—

"Here I mark one thing, viz.: That while this wire, with those other co-related telegraphs on either side, will, in the first instance, work towards monopoly, in the second and main instance they will work towards diffusion and the common weal; for although commerce and politics, and the merchant class and the political class will in the first instance be the users, and so be benefited first by it, yet, in the main, the people will be the ones who will reap the benefit, for whatever thing brings now whole communities into circumstances of greater prosperity must needs, in the spirit and temper of our time, be distributive. If it were possible for knowledge to be confined to the minds of the many, if it were possible for monopolists to lock up the two ends of this wire, holding, on either side, the power of its intelligence, it might be disastrous to governments and to the people. But now it has a tendency to make knowledge co-extensive with the globe instantaneously."

Illuminations took place in Brooklyn, the city of Captain Hudson's residence. Among the transparencies was one hung over the Mechanics' Bank, which bore the words of a telegraphic dispatch forwarded by the commander of the *Niagara* to his family—viz.:—

"Trinity Bay, Aug. 5, 1858.

"God has been with us. The Telegraph Cable is laid without accident, and to Him be all the glory. We are all well.

"Yours affectionately, Wm. L. Hudson."

Similar demonstrations occurred on the 5th and 6th of August at the South and West. The cities of Philadelphia and New Orleans were especially jubilant. In the city of Boston, salutes were fired, illuminations took place, and bonfires were lighted.

The reception of the Queen's Message on the 16th and 17th of August, however, produced the most remarkable effect. Celebrations which had been promised, but were postponed until the certainty of success became assured, took place under circumstances of unusual impressiveness. That which occurred in the city of New York, on the evening of Tuesday, August 17, was in many respects the most remarkable popular demonstration that has occurred for many years. Without preparation, and without the settlement of a programme, the city suddenly exulted. It was a momentary impulse, ended as speedily as it began, and, being a short-lived excitement, brought no reaction with it. Two thousand workmen from the Central Park appeared in Broadway at noon, marching in orderly procession, bringing with them their implements of labor which they had ceased using for a day. From the City Hall, the hotels and offices, flags floated. Banners were displayed in the leading thoroughfares, as night drew on; and at dusk the City Hall, the Astor House, the newspaper establishments, stores and dwellings in the lower part of the city blazed with the light of a spontaneous illumination. Unfortunately a display of fine pyrotechny which took place in front of the City Hall, resulted in a disaster to that fine edifice; the unconsumed remains of the fireworks igniting the roof of the Hall and producing a partial destruction of the upper floor. The celebration was marked by the burning of a City Hall, and the two events became historical together.

Among the devices displayed upon transparencies during the illumination, the following expressed the popular feeling in a significant manner:—

"VICTORIA. All hail to the Inventive Genius and Indefatigable Enterprise of JOHN and JONATHAN, that has succeeded in consummating the Mightiest Work of the Age; may the Cord that binds them in the bonds of International Friendship never be severed, and the FIELD of its Usefulness extend to every part of the Earth.

"Let nations shout, 'midst cannons' roar,
Proclaim the event from shore to shore."

"The old CYRUS and the new—the first conquered the land for himself, the second the ocean for the world."

"Lightning Caught and Tamed by FRANKLIN. Taught to Read and Write and go on errands by MORSE. Started in the Foreign Trade by FIELD, COOPER & Co., with JOHNNY BULL and BROTHER JONATHAN as Special Partners."

In the city of Brooklyn, on the succeeding evening, a jubilee reception was given to Captain HUDSON. In response to an address of welcome from ex-Mayor HALL, Captain HUDSON made the following pithy little speech:—

"MR. MAYOR: If I were not a sailor, I should stay back and not open my mouth to reply. But I am surprised at what I see before me; for I have done nothing to deserve this—it is entirely undeserved. I have done nothing more than my duty—nothing that calls for anything of this kind. We have been engaged in a work that will make a sensation over the whole world. It ushers in a new era in the commercial community. Through its means we shall be able to carry the gospel to all parts of the earth. We have been simple instruments in Higher Hands to bring about the accomplishment of this enterprise. There is One higher than us, and to Him should be given all the praise. I hardly know how to speak. But you would not expect much from a sailor. and therefore you will not be disappointed."

The jubilee was general throughout the country, after the reception of the Message of Her Majesty; so general, indeed, were the demonstrations of unbounded joy, that it would be difficult to exaggerate any account of the rejoicings. The jubilee of a Nation is a rare event, and justifies a passing notice in this record.

At Belfast, Maine, the celebration was very spirited. H. O. ALDEN, Esq., Vice-President of the American Telegraph Company, and a stockholder in the Atlantic Company, caused the Queen's Message to be generally circulated among the citizens.

At Detroit, Michigan, the display of enthusiasm was totally unprecedented. The demonstrations commenced by the firing of one hundred guns at sunset; fire-bells rang merry peals, rockets were fired, bonfires lighted in the crowded streets, the thoroughfares gaily decorated with streamers and appropriate transparencies. The public buildings and private dwellings were illuminated. An imposing torch-light procession, headed by the Mayor of the city and officers, marched through the principal streets to the *Campus Martius*, where addresses were delivered, and expressive resolutions passed.

The city of Cincinnati, Ohio, was brilliantly illuminated in honor of the Ocean Telegraph. At the corner of every street barrels of tar were burning, donated for the purpose by the Gas Company. The bells of Cincinnati, Covington, and Newport were rung, and 100 guns were fired.

At Pittsburg, Pennsylvania, the celebration was a brilliant affair. At four o'clock, P.M., all business was suspended, all the bells in the city were rung, and a salute of 100 guns fired. In the evening there was a torchlight procession by the civic societies, and a general illumination of public and private houses. On the rivers there was an illuminated regatta by all the boat clubs.

The reception of the Queen's and President's Messages was hailed at St. Louis with delight. The Messages were read on 'Change, and elicited hearty acclamations. Congratulatory remarks were made by several gentlemen. A meeting was convened under the direction of the President of the Chamber of Commerce, and the following resolution unanimously adopted:—

Resolved, By the Chamber of Commerce, that St. Louis will unite with the country in the celebration of this great international enterprise.

X.

MR. BERDAN AS THE INVENTOR OF THE NEW PAYING-OUT MACHINE.

[THE name of Mr. HIRAM BERDAN, of New York, having been mentioned in the body of this work, in connexion with the invention of the Paying-out Machine, which was successfully employed in laying the Atlantic Cable in the Summer of 1858, we deem it proper to produce certain proofs, establishing the claims of that gentleman as the original inventor of that beautiful apparatus :]

Soon after the failure of the Telegraphic Expedition of 1857, and as soon as the causes of such failure could be satisfactorily ascertained, Mr. H. BERDAN, of this city, set himself to the task of investigating the subject and devising a paying-out apparatus which would supply the defects of that already used, and in his judgment, successfully lay the Cable across the Atlantic.

About the 1st of January of the present year, Mr. BERDAN had completed his design and had a working model constructed on such a scale as to show clearly the full operation of all the different parts of the apparatus. Mr. FIELD's attention had been called to Mr. BERDAN's invention, and expressing his anxious desire that Mr. EVERETT should examine the apparatus, a day was appointed for him to do so, which was the Monday previous to the sailing of the Persia, which took Mr. FIELD and Mr. EVERETT to Liverpool in January last. Mr. FIELD, Mr. EVERETT, PETER COOPER, Capt. HUDSON, and many other gentlemen, were present when Mr. EVERETT made his first visit to, and examination of, Mr. BERDAN's model.

When Mr. EVERETT first examined the model and noticed that the tension on the Cable was to be tested by a counter weight, he remarked with reference to it, "Oh, that won't do, a counter weight cannot be used on board ship, the momentum is too great. We have thought of that and abandoned it."

To which Mr. BERDAN replied, that he doubted if Mr. EVERETT yet understood the operation of the machine and its effect upon the Cable to give elasticity to it; that there was no serious difficulty in placing the counter weights on board ship. Mr. BERDAN then operated the model, showing the effect of the counter weights and the various parts of the machinery so satisfactorily that Mr. EVERETT appeared to change his mind at once, and observed, "I came here with no expectations of seeing any thing new. We have had a thousand suggestions from various sources, but this is new and interesting, and I am now satisfied that we must adopt a compensating movement." Mr. EVERETT then remarked that he had no plan of his own as yet, was wedded to no system, and was determined to seek for the best, and use it; that he was much pleased with Mr. BERDAN'S plan; that he wished Mr. B. would let him take the model with him that he might study it on his way out to London. Mr. BERDAN consented, and the model was accordingly boxed and sent to Mr. EVERETT'S state-room, on board the Persia, and was taken by Mr. FIELD and EVERETT to London. The following is a copy of a letter acknowledging the receipt of the model by the Atlantic Telegraph Company:

"ATLANTIC TELEGRAPH COMPANY,
"22 Old Bond street, London,
"ENGINEERS' DEPARTMENT, *March 4th*, 1858.

CYRUS W. FIELD, Esq., Atlantic Telegraph Company, London:

"MY DEAR SIR,—I have read the very interesting communication communicated by Mr. BERDAN, of New York, addressed to yourself, and have also examined the model of his apparatus which you brought from America. I must say that he deserves our very best thanks for the elaborate attention he has devoted to the subject which has resulted in the design of a paying-out machine, embodying many of the conditions requisite for success.

"The most important feature of his plan is the arrangement to compensate for the motion of the vessel, and this you are aware has engaged our attention for some time."

On the 25th of January last, Mr. BERDAN forwarded to Mr. FIELD drawings of his paying-out apparatus, accompanied by a letter giving his views generally, and full explanations of his machine in particular.

This communication from Mr. BERDAN to Mr. FIELD was laid before Mr. BRIGHT, the Engineer-in-chief of the Company, who wrote to Mr. FIELD in reply, of which the following is an extract:

"THE ATLANTIC TELEGRAPH COMPANY LIMITED,
"22 Old Bond street,
"LONDON, E. C., *January* 19*th*, 1858.

"SIR,—I have received, by the hands of CYRUS W. FIELD, Esq., one of the Directors of this Company, your very beautiful and elaborate model of a paying-out machine for depositing the Atlantic Telegraph Cable in the Ocean.

"I beg to tender you the warmest thanks of the Directors for the lively interest you have so evidently evinced for the success of this undertaking.

"The machine shall be submitted to the English and American scientific authorities who are engaged in assisting us with their judgment as to the right means for laying the Cable next summer, and I am quite certain it will receive at their hands a perfectly fair and especial consideration. Again thanking you for your kindness,

"I am, Sir, yours very truly,

"GEO. SEWARD, *Secretary*.

"HIRAM BERDAN, Esq., 110 Broadway, New York."

It is pertinent to remark that in Mr. Berdan's letter to Mr. Field of the 25th of January, the former laid great stress upon the importance of properly adjusted scrapers to remove the tar from the sheaves. In that letter, in describing his machinery, Mr. Berdan writes:—"At a convenient point in the frames B and D, as well as the other sheave frames, I place shafts, having scrapers attached thereto, of such forms as to fit the grooves in the sheave wheels. The axes of these shafts are placed at such a distance from the periphery of the sheave wheels as to allow the shaft to rotate freely on its axis, the scrapers just touching the whole inner surface of the grooves in the sheaves; a handle on the outer end of the shaft serves to hold the scrapers in a proper position to remove any tar that may be adhering to the surface of the sheaves. On the top of the frames, and immediately over the sheaves, I place water cans, having an outlet over each sheave to wet the Cable, and prevent in part the accumulation of tar on the sheaves, as well as to assist in the cleaning operation of the scrapers." Again, in Mr. Berdan's letter to Mr. Field, dated the 27th of March last, Mr. Berdan writes as follows:—"Constant dripping of water on the sheaves, together with the use of an instrument which can be applied with accuracy every few moments to take off the accumulation of tar, is also very important,—water is preferable to oil, as the latter would dissolve the tar on the Cable, which is not desirable."

On the 3d of June following, Mr. Everett makes a report to Mr. Field in which occurs the following:—"The operation of the machinery generally, is certainly satisfactory, and there is no alteration I can suggest other than in the tar scrapers, which will require modification; the amount of tar accumulating is so much beyond what could have been expected from last year's experience, owing to the repeated coatings since it was unloaded from the vessel last October, that extraordinary provision will be required." It clearly appears from Mr. Everett's report that Mr. Berdan did not attach too much importance to the tar scrapers, and that his recommendations, which ought to have been adopted at first, were found absolutely necessary at last.

The question now accrues, and it is one of importance: To whom is the credit due for devising the paying out apparatus which was used in successfully laying the Atlantic Cable? Mr. Everett is doubtless entitled to a full share of credit for superintending the constructing of the machinery, and for the exercise of a sound discretion in adopting from the numerous different suggestions submitted to the Company, such as appeared to him most feasible. It seems equally clear that in the exercise of his judgment in converse with the other able engineers of the Company, the principle of Mr. Berdan's compensating apparatus, and the arrangement and use of the tar scrapers, with some modifications, were adopted and used; and as these two features in the apparatus were the principal novelties and improvements upon the old machine, it is certainly fair to give Mr. Berdan the full credit of their introduction, especially as Mr. Everett and the other engineers had the model, drawing, and specifications of the same constantly before them.

CATALOGUE

OF THE

PUBLICATIONS

OF

RUDD & CARLETON,

310 Broadway,

New York.

TRUE LOVE NEVER DID RUN SMOOTH.

An Eastern Tale, in Verse. By THOMAS BAILEY ALDRICH, author of "Babie Bell, and other Poems." Elegantly printed, and bound in muslin, price 50 cents.

DEAR EXPERIENCE.

A Tale. By G. RUFFINI, author of "Doctor Antonio," "Lorenzo Benoni," &c. With illustrations by Leech, of the *London Punch*. Muslin, price $1 00.

FOLLOWING THE DRUM;

Or, Glimpses of Frontier Life. Being brilliant Sketches of Recruiting Incidents on the Rio Grande, &c. By Mrs. EGBERT L. VIELÉ. Muslin, price $1 00.

COSMOGONY;

Or, the Mysteries of Creation. A remarkable book, being an Analysis of the First Chapter of Genesis. By THOMAS A. DAVIES. Octavo, muslin, price $2 00.

NOTHING TO WEAR.

A Satirical Poem. By WILLIAM ALLEN BUTLER. Profusely and elegantly embellished with fine Illustrations by Augustus Hoppin. Muslin, price 50 cents.

THE SPUYTENDEVIL CHRONICLE.

A brilliant Novel of Fashionable Life in New York. A Saratoga Season Flirtations, &c. A companion to the "Potiphar Papers." Muslin, price 75 cents.

BROWN'S CARPENTER'S ASSISTANT.

The best practical work on Architecture, with Plans for every description of Building. Illustrated with over 200 Plates. Strongly bound in leather, price $5 00.

THE ATLANTIC TELEGRAPH.

An Authentic History of that great work; with Biographies, Maps, steel and wood engravings, Portraits, &c. Dedicated to CYRUS W. FIELD. Muslin, price $1 00.

ISABELLA ORSINI.

A new and brilliant novel. By F. D. GUERRAZZI, author of "Beatrice Cenci;" translated by MONTI, of Harvard College. With steel portrait. Muslin, price $1 25.

K. N. PEPPER PAPERS.

Containing Verses and Miscellaneous Writings of one of the first humorous contributors to "*Knickerbocker Magazine*." With Illustrations. Muslin, price $1 00.

THE AMERICAN CHESS BOOK.

Prepared by PAUL MORPHY, LOUIS PAULSEN, and the chief Members of the late New York Chess Congress for 1857. Muslin, 12mo. Fully illustrated. (*In press*).

THE CAPTIVE NIGHTINGALE,

And other Tales. Translated from the German of KRUMACHER. A charming book for the Young. Fully illustrated. Muslin, gilt back, &c., price 50 cents.

STORIES FOR CHILDHOOD.

By AUNT HATTY (Mrs. Coleman). Beautifully bound in cloth gilt, and profusely illustrated. Put up in boxes containing 12 assorted volumes. Price per box, $4 00.

GOOD CHILDREN'S LIBRARY.

By UNCLE THOMAS. A dozen charming stories, beautifully illustrated; bound in cloth, gilt backs. Put up in boxes containing 12 assorted volumes. Price per box, $4 00.

HUSBAND *vs.* WIFE.

A Domestic Satirical Novel. By Henry Clapp, Jr. Illustrated by A. Hoppin, in colors, on cream paper, in Illuminated Missal style. Muslin, price 60 cents.

ASPENWOLD.

An Original Novel, of interest and merit. By an American writer. Illustrated with original designs on wood by Howard. Muslin, price $1 25.

LECTURES OF LOLA MONTEZ,

Including her "Autobiography," "Wits and Women of Paris," "Comic Aspect of Love," "Beautiful Women," "Gallantry," &c. Muslin, steel portrait, price $1 00.

LIFE OF HUGH MILLER,

Author of "Schools and Schoolmasters," "Old Red Sandstone," &c. From the Glasgow edition. Prepared by Thomas N. Brown. Muslin, price $1 00.

KNAVES AND FOOLS;

Or, Friends of Bohemia. A Satirical Novel of Literary and Artistic Life in London. By E. M. Whitty, of the *London Times*. Illustrated. Muslin, price $1 25.

THE COTTAGE COOK BOOK;

Or, Housekeeping made Easy and Economical in all its Departments. A most practical and useful work. By Emily Thornwell. Muslin, price 75 cents.

THE CHRISTMAS TREE.

A valuable Book of Amusement and Instruction for the Young, containing Stories upon every subject. Illustrated by Kenny Meadows. Muslin, price 75 cents.

EROS AND ANTEROS;

Or, the Bachelor's Ward. One of the very best of Modern Novels. A story of strong interest, and written with true poetic feeling. Muslin, price $1 00.

DOESTICKS' LETTERS.

Being a compilation of the Original Letters of Q. K. P. Doesticks, P. B. With many comic tinted illustrations, by John McLenan. Muslin, price $1 00.

PLU-RI-BUS-TAH.

A song that's by-no-author. *Not* a parody on "Hiawatha." By Doesticks. With 150 humorous illustrations by McLenan. Muslin, price $1 00.

THE ELEPHANT CLUB.

An irresistibly droll volume. By Doesticks, assisted by Knight Russ Ockside, M.D. One of his best works. Profusely illustrated by McLenan. Muslin, price $1 00.

NOTHING TO SAY.

A Satire in Verse, which has "Nothing to Do" with "Nothing to Wear." By Doesticks, P. B. With Illustrations by McLenan. Muslin, price 50 cents.

THE OLD LOVE AND THE NEW.

A very charming Novel, containing the very elements ot success. Written by Mrs. Juliet H. L. Campbell. Handsomely bound in muslin, 12mo., price $1 00.

LIFE OF SPENCER H. CONE.

Being Memoirs of the late Pastor of the First Baptist Church in the city of New York. Prepared by his Sons. Steel Portrait. Muslin, price $1 25.

www.ingramcontent.com/pod-product-compliance
Lightning Source LLC
LaVergne TN
LVHW010243110826
845151LV00004B/1381

9781425524296